IMAGES OF ENGLAND

COLLIERIES OF NORTH EAST LANCASHIRE

There were originally two shafts at Bank Hall Colliery, the No.1 and No.2 Shaft, which was always referred to as the 'Arley Shaft', and is seen here with the later No.3 Shaft, or 'Dandy Shaft'. The Arley (40ft below the Dandy) was the best seam ever worked in the Burnley Coalfield. A coal seam in this part of Lancashire was always called a mine. Between the two shafts can be seen the cupola, a ventilation chimney which was sunk down to the Dandy mine, inset in the shaft. A furnace provided a powerful draught up the chimney, causing fresh air to be drawn into the workings of the mine and driving out the gasses and foul air, a method of ventilation common at the time. The No.2 shaft remained in full use until 1932, when winding from the Arley mine ceased at Bank Hall. Coal-winding on a reduced scale was carried out here until 1939, for all the Arley coal from Reedeley Colliery and Wood End Colliery workings was raised here along with what was named the 'Barden Chain Road'. In the 1890s, Bank Hall employed 340 men underground and 120 surface workers. (W. Rawstron Collection)

IMAGES OF ENGLAND

COLLIERIES OF NORTH EAST LANCASHIRE

JACK NADIN

The Author

Jack Nadin worked at North East Lancashire's last deep coal mine, the Hapton Valley Colliery, from 1964 through to 1971. He comes from a line of miners; his grandfather worked at Hapton Valley, as did his uncles. He has a passion for anything to do with local history in his home town of Burnley, Lancashire – especially that relating to the coalmining industry and the industrial archaeology of the area, and is a regular contributor of local history articles to newspapers and magazines in and around Burnley.

Titles published by the author on coalmining history include: *Hapton Valley Colliery: The History of an East Lancashire Colliery*, published by the Burnley and District Historical Society; *The Coal Mines of East Lancashire* and *Coal Mines around Accrington and Blackburn*, both published by the Northern Mines Research Society; *Happy Valley No More: The History of Hapton Valley Colliery*; and *Bank Hall Colliery: Burnley's Grandest Pit*. His titles on the general local history of the Burnley area include *Old Ightenhill*, *Old Rosegrove* and *Old Lowerhouse*, sub-districts of Burnley, and *Burnley's Industrial Heritage: The Cotton Mills*. He has reprinted a number of historical booklets on Burnley's past, including *Bygone Burnley*, a series of articles published in the *Burnley News* from 1933, and *Rambles around Burnley and Padiham*, first published by Smith Fielding in 1910.

First published in 2003 by Tempus Publishing

Reprinted in 2010 by
The History Press
The Mill, Brimscombe Port,
Stroud, Gloucestershire, GL5 2QG
www.thehistorypress.co.uk

Reprinted 2013

© Jack Nadin, 2010

The right of Jack Nadin to be identified as the Author
of this work has been asserted in accordance with the
Copyrights, Designs and Patents Act 1988.

All rights reserved. No part of this book may be reprinted
or reproduced or utilised in any form or by any electronic,
mechanical or other means, now known or hereafter invented,
including photocopying and recording, or in any information
storage or retrieval system, without the permission in writing
from the Publishers.

British Library Cataloguing in Publication Data.
A catalogue record for this book is available from the British Library.

ISBN 978 0 7524 2803 1

Typesetting and origination by Tempus Publishing Limited
Printed and bound in England.

Contents

Acknowledgements 6

Introduction 7

1. The Colliery Masters 9
2. The Collieries: Early Views 11
3. Nationalisation: Out with the Old, In with the New 44
4. The Ginney Systems Around Burnley 87
5. People and Places 91
6. Sporting Days 113
7. The Private Coal Mines 120
8. The End of Coalmining in North East Lancashire 124

Acknowledgements

Many of the photographs illustrated within this book have never before been published. Some are taken from two large albums held at Burnley Central Library and were donated to the reference library by former Bank Hall Colliery manager W.E. Rawstron. They were used to illustrate a paper read on 18 March 1958 about the reorganisation of the local mining industry between 1947 and 1957. Sadly, William (Bill) Rawstron died in July 1984. I was more than delighted, however, when his son, Ian Rawstron, gave me permission to use the photographs from his father's collection to illustrate this book. Many of these photographs are now over fifty years old and, without his permission, these images of the pits in the Burnley area would in all probability have lain confined to a backroom shelf in the reference library. I for one am eternally grateful – thank you, Ian.

Photographs from others who contributed are acknowledged in the text – to those too, a big thank you. Without these contributions, this pictorial account of Burnley's mining past would not have been possible.

As well as thanking Tempus Publishing for producing this book, I would like to give special thanks to the staff at Burnley Reference Library and Accrington Reference Library for their time and knowledge. Thanks also to Ken Spencer; Steven Spencer; Harry O'Neill; Clive Seal; Gordon Hartley; Peter Nadin and John Simpson Little. I must also mention the many members of the public who responded to my appeals for names and information on old photographs published in the local newspapers, the *Burnley Express* and the *Lancashire Evening Telegraph*, and thanks to these for publishing those appeals.

Lastly, but not least, special thanks to my wife Rita for all her patience while I was undertaking research and for the many cups of coffee while I was endlessly stuck in front of the computer monitor typing away

Introduction

The Burnley Coalfield covers an area of Lancashire from Colne in the north through to Blackburn and Darwen in the south, Higham to the west and Worsthorne Moor to the east. A number of pits in the Bacup area have also been included as they were listed along with the Burnley Pits under the National Coal Board's (NCB) N.W. Division, No.4 Area – Burnley. Whilst not as large as many of its counterparts throughout Lancashire, such as the Wigan Coalfield, the area was nevertheless worked for coal from the earliest times. However, the maximum number of the larger collieries that ever worked here at one time never exceeded two dozen, and it is these pits we shall be looking at.

Early references to mining in the area are common. In 1294, for instance, the monks of Bolton Priory in Yorkshire were obtaining 'sea coles' from a mine at Trawden, Lancashire, for use in their smiths' forges. It is in this area that the seams of the Burnley Coalfield 'outcrop', or rise to or near the surface. These early pits would have been shallow workings, with the coal being dug out or 'outcropped' where it appeared at the surface.

Coal mines are mentioned in the *Manor of Colne* for 1311 and 1323, while in 1423 mines were leased to Edmund Parker in the Forest of Trawden for 13s 4d per year. At Clayton-le-Moors, in the west of the Burnley Coalfield, coal was mined as early as 1376, and in 1529, during the reign of Henry VIII, coal from a mine in Padiham was used as fuel at Whalley Abbey. In around 1450 two brothers, Thomas and William Watmough, found coal while digging for ironstone on Broadhead Moor (Burnley), which they began to mine and sell to meet the expenses of their mining operations. It was here that the first 'turne' or 'windglasse' (windlass) was erected. The mine produced over 240 tons of coal in 1526, which was sold at 3d per ton. Eighty years later, the coal from Broadhead Moor Pit was being sold at 1d per horse load, or as much as could be carried in the panniers, baskets or sacks on the animal's back.

From around the early seventeenth century, a number of coal pit shafts were sunk by the Shuttleworth family of Gawthorpe Hall, near Padiham. These shafts were circular in shape, about 6ft in diameter, and stone lined. One of these was positioned close by Brookfoot Farm, Padiham, and the coal was worked by the pillar and stall method. The area around Grove Lane has been extensively worked for coal in the past and today the many hollows in Grove Wood still indicate the location of old shafts and mine workings.

Even the Towneley family of Towneley Hall, Burnley, got involved in mining coal. Following the Civil War (1642-1649) a Towneley Estates report indicated that there was a mine of the 'shaft type' at Cliviger. The Towneley family had in fact leased a coal mine in Cliviger to William and Sarah Cockcroft for a thirty-one-year term for the sum of £5, together with an annual rent of 5s. The Cockcroft's, however, claimed a reduction because they had spent £350 sinking the pit but had been 'much impeded by the brittleness of the earth and the overflow of water'. The Towneley family also had a coal mine at Hapton on lands belonging to Hapton Hall Farm, and Hapton Clough was sunk by Charles Towneley in 1779.

The Hambledon Hill Colliery was first developed as a drift mine in 1797 by Harry Winterbottom of Goodshaw Chapel, near Rawtenstall, the remains of which can still be traced today. In later years the Cupola Pit Colliery at Hambledon became part of the Hambledon Hill Colliery and was used as a furnace ventilation drift. Drawn-up contracts stated 'that the lessees were to be responsible for filling in and taking away any surplus waste such as shale or slack, when the reserves of the mines had been exhausted'.

Coalmining was introduced among the Burnley industries between 1500-1650, and it was during this period that the mineral wealth of the town was developed. In these early days, only small amounts of coal were worked, and these at the outcrop of the seams. At this time there were two coal pits leased at Brunshaw, a pit at Habergham, and two pits on the Ridge at the former Rappock Lane. The Halmot courts fixed a fine of 3s 4d per person 'Out of Craven' (from Yorkshire) who were found digging for coal in Colne township. At this time the Tume Hill area of Colne was also being worked for coal, and coal around 'High Marsden' was being sent to Yorkshire including, among other places, Fountains Abbey.

Various seams of coal were worked in the area, including the 10/11ft-thick 'Padiham Thick Seam', located between Dean Wood, to the west of Padiham, and Pendle Hall Farm, to the east. In the mid-seventeenth century a small coal mine existed at the junction of Grove Lane and Garden Street which worked the Padiham Thick Seam. What was long thought to have been a horse drinking trough in Slade Lane, Padiham, was, in fact, the entrance to an old drift mine that led to the Padiham Thick Seam. Another old drift mouthing was located around 20yds to the south. At the turn of the twentieth century the area had been affected by Foot and Mouth and large cavities that had been mined in the Padiham Thick Seam were used to dispose of the dead animals which, once placed in the holes, were set fire to. When the whole operation was completed the holes were back-filled, leaving no sign of the cremation until the seams were once again exposed during the opencast operations.

This area had long been worked for coal by the locals and farmers to supply their own needs, so much so that very little coal was actually bought in Padiham, and indeed some of it was even sold on.

Other ancient workings include those at Caster Cliffe, near Nelson, and those at Higham. Those at Higham can be seen north of the church on Brown Hill, where subsidence in 1966 exposed a 95ft-deep shaft, below which all was flooded. Two other shafts can be seen here, represented on the surface by old pit heaps arranged around the shafts – an early form of bell pit working. The outcrop of the Arley Mine lies in this area, a mine which reputedly produced the best coal in the Burnley Coalfield. In 1777 the cost of the best Arley coal was 7d for 120lbs. Coal mines were also worked on the Huntroyde Estate at Padiham, where old wooden shovels have been found tipped with iron to give better service, along with an old pair of wooden clogs. Examples of these wooden tools can be seen at Towneley Hall. As for winding the coal to the surface, horse whims or gins (an abbreviation of 'engine'), are known to have been used in the Burnley Coalfield at Ightenhill, the old Pendle Forest Pit at Wheatley Lane, Cliviger, Padiham and Oswaldtwistle.

The nineteenth century saw child labour in the mines, when children aged five, six or seven years old were taken underground. Sixteen children between the ages of ten and fifteen worked at the Ightenhill Park Pit at Burnley during the 1840s. Happily, there appears to have been no widespread ill treatment of the children in the Burnley area, as was common elsewhere. Indeed the local coal masters, such as the Thursbys, even provided schools for the children, long before ordinary working children had such opportunities.

It was at this time, during the great Industrial Revolution, that the larger coal mines were developed in the Burnley Coalfield, and these were soon dominated by four companies: John Hargreaves, George Hargreaves, Brooks & Pickup and the Cliviger Coal Co. These are the companies that sunk the pits which many of the Burnley miners will, even now, recall, and this is the history of those pits, of those who sunk and financed them, and of those who toiled, laboured, even perished, within their depths. Save for one or two exceptions, as we shall see, the Burnley Coalfield never suffered the effects of the great colliery explosions that occurred from the 1850s through to the 1900s in the Lancashire Coalfields. However, there were many singular accidents which collectively claimed many more lives than all these disasters put together. The price of coal was indeed high, not only in the Burnley Coalfield, but throughout Britain – something we should never forget.

When the NCB took over the local collieries in 1947, there were nineteen pits in the North Western Division of what became the No.4 Area. Between them these pits employed 3,443 men, both underground and on the surface. The coal mines of North East Lancashire have now gone, as have most throughout the country, and with their demise went a comradeship and companionship that few other industries enjoyed. The last deep mine in the Burnley Coalfield, Hapton Valley Colliery, closed in 1982.

One
The Colliery Masters

The Exors of Col. John Hargreaves were by far the largest colliery proprietors in the Burnley Coalfield, and with this in mind it would be wise to give a brief history of the firm.

The coal industry in the early nineteenth century consisted of a number of small mines used to supply local demand as fuel for the home fire and the odd mill or foundry. By the 1850s, new and deeper collieries were being sunk to cope with the demand for the hungry mills in East Lancashire during the Industrial Revolution. Burnley was at this time famed for its 'Three Cs': cotton, coil (coal) and clogs. However, to sink a brand new colliery needed huge investments; money that would be slow in returning. Even the local and landed gentry could ill afford to take such risks as sinking new pits, and this was left to the soundly established firm of the Exors of John Hargreaves, later Hargreaves Colliery Co., a company that owed its existence to the marriage, in 1755, of the Revd John Hargreaves to the widow of Henry Blackmore, who owned the colliery at Fulledge.

The son of James Hargreaves and Elizabeth Birtwistle, the Revd Hargreaves was born in 1732 and died in 1812 at Bank Hall, Burnley. In 1797 he bought the leaseholds of practically all the mineral rights in Burnley. The Revd John Hargreaves and his wife died without issue, and their estate went to his brother's children, James Hargreaves and Col. John Hargreaves.

When Col. John Hargreaves (1775-1834) married Charlotte Ann Ormerod in 1802 he gained the wealth and lands of the Ormerod estates. The couple had three children, John, Eleanor Mary and Charlotte Ann. Charlotte Ann married James Yorke Scarlett (of the Charge of the Light Brigade fame) but they died without issue. The other daughter, Eleanor Mary, married the Revd William Thursby, and so the Hargreaves Collieries passed into the hands of the Thursby family.

On 1 April 1834, on the death of Col. Hargreaves, the company was named 'The Executors (Exors for short) of John Hargreaves Collieries Co.' This company, which later became the Hargreaves Colliery Co. in 1932, dominated all the Burnley area mining activities up to the Nationalisation of the British coalmining industry in 1947, sinking many of the pits mentioned in the text. On 1 January 1932 it was announced that arrangements were to be made for the amalgamation of the Burnley, Accrington and Altham Collieries of the Exors of John Hargreaves Ltd, George Hargreaves & Co. Ltd, and Altham Colliery Co. Ltd (1924). The merged company was given the name Hargreaves Collieries Ltd.

A group of men and boys at the Cliviger Coal Co.'s Railway Colliery, c.1920. The leases for these mines were drawn up between William Edmondson, said to have been the father of Cliviger Coal, and J. Collinge, J. Haigh and H. Clegg in 1813 and 1823. Four mines were working at Cliviger in 1822: the Bankwell Level, Broughtons Pit, the Turner Carr Level, and the Overtown Pit in the village. However, these were small ventures and by the 1840s work was concentrated on the Railway Pit and at Copy Bottom, Copy Colliery. By the 1850s, demand for coal from the growing cotton trade and for household and steam coal was outstripping supply, so the company sunk the Union Colliery, which was brought into operation in 1855. By the early 1870s a gasworks was erected behind the colliery, and this supplied many of the pit workers' houses with gas right up to the closure of the Union Pit in 1942. Railway Pit survived until around 1938. Only Copy Colliery made it into Nationalisation, and this too was closed down and salvaged out in March 1964, the last of the former Cliviger Coal Co.'s pits.

Thomas Brooks of Brooks & Pickup. The company worked the Towneley Colliery at Burnley under the titles of Brooks & Pickup; Brooks & Brooks Collieries Ltd; and later the Towneley Coal & Fireclay Co. Ltd. Peter Pickup, born c.1829, was the junior member of this partnership. Around 1841 he joined into a partnership with Thomas Brooks and thus Brooks & Pickup colliery proprietors was born. Their first ventures together included the Wholaw Nook Colliery near Clowbridge Reservoir, and the Cupola Colliery at Hambledon Hill. Brooks & Pickup sank the Towneley Colliery in 1869 and also worked the Waterloo Main Colliery at Leeds, which employed over 800 men in the 1890s. Following the death of Peter Pickup, the firm went under the title of Brooks & Brooks, and later Towneley Coal & Fireclay Co. Ltd. The business was worked by Thomas's sons, William and Gerald, and others, following the death of their father. Their main pit in Burnley, Towneley Colliery, was closed down soon after Nationalisation.

Two

The Collieries: Early Views

Bank Hall Colliery, sunk in 1865 by the Exors of John Hargreaves, later Hargreaves Collieries, was the largest working pit in the Burnley Coalfield. This 1903 view shows the No.1 shaft at the colliery, with a Clayton-Goodfellow Pumping Engine dated 1886. The L leg converted the rotary motion of the steam engine into the up and down motion needed to work the pumps in the shaft. The pumping engine had two horizontal cylinders, 34in by 60in, connected by a wrought iron horizontal rod to the large L leg placed over the shaft. The engine raised water in three-bucket lifts from a depth of 240yds. The 18in-diameter buckets had a stroke of 7ft, and the highest lift delivered 77 gallons per stroke to the surface. The engine usually worked eighteen hours in twenty-four, at four strokes a minute. Notice too the ginney – developed by a member of the Landless family at Marsden Colliery – with its endless chain, a system of haulage used extensively both underground and on the surface in the Burnley area mines. The colliery was sunk by the Exors of John Hargreaves and owned by Hargreaves Collieries till Nationalisation in 1947. In 1945 the colliery, managed by W.E. Rawstron, employed 571 men underground and 362 surface men. (W. Rawstron Collection)

The No.3 Shaft at Bank Hall was sunk in 1903. This larger-diameter shaft was sunk 186yds to the Dandy Mine, and became known as the 'Dandy Shaft', colloquially the 'King and Dandy Shaft', for the King Seam was also worked from this shaft. The No.3 Shaft, a coal-winding shaft, was also utilised as a downcast shaft in ventilation. The return air from the King and Dandy workings was routed around the existing workings through to the cupola furnace shaft and vented into the atmosphere. After the cupola shaft ceased to exist in later years, the return air was routed to the No.1 shaft, which later became the upcast shaft for the whole of Bank Hall until closure. Here, in August 1948, we see the Dandy Shaft after being filled and the headgear being dismantled. (W. Rawstron Collection)

This scene at Bank Hall Colliery shows the ginnies from Clifton Pit, the Arley Pit at Bank Hall itself, and the area where the coal from Bee-Hole and Rowley Collieries arrived. This was close by to the canal basin. Coal was also raised at Bank Hall from Reedley Pit, and it too was screened here. (W. Rawstron Collection)

The No.1 Pit, upcast shaft and Arley Pit at Bank Hall Colliery, with offices, lamp room and workshops in September 1948. The pit is obviously outdated at this time, with old wooden buildings and derelict sheds being passed off as the general surface layout of the pit. Little was invested under the old owners, Hargreaves Collieries, even though the pit was still very much a viable operation. Very soon after this photograph was taken the pit was upgraded under the NCB in a bid to beat the dismal coal shortage at the time – a project that was to cost over £1 million. (W. Rawstron Collection)

Above, left: The No.4 Shaft at Bank Hall Colliery from the Arley Pit Bank, September 1948, showing the ginney, the screens and the colliery lamp room. Just after Nationalisation, the colliery employed 605 men underground and 412 surface workers. Mr Rawstron, the manager, had the foresight of collecting many of the photographs in this book, which he used on a number of occasions in his talks on the local mining industry. Today, Bank Hall Park sits on the site of all this activity in days gone by. (W. Rawstron Collection)

Above, right: This sketch, the only known image of Bee-Hole Colliery, was taken from a *Burnley Express* report on a football match between Burnley and Blackpool held on 10 February 1892 – Burnley won! This end of Burnley's Turf Moor football ground was, even in recent years, known as the 'Bee-Hole End'. In the background can be seen the colliery's two shafts, winding engine room and chimney. The two headgears at the colliery were apparently made of wood. The first of the Bee-Hole Pits was allegedly sunk by Henry Blackmore, who also sank the nearby Fulledge Pit, and dated from around the 1750s, perhaps even earlier. The second Bee-Hole Pit, seen here, was sunk in 1872. The pit worked the pillar and stall system of mining and, because the seams were only 35yds deep, it was said that when the miners were having their bait (lunch), they could hear the weavers and looms overhead in Park View Shed. There were no baths at Bee-Hole and the miners sometimes had to go home in wet clothes, especially when they had to work in water for an extra shilling (5p) a day. The Bing Seam was also worked at the Bee-Hole in the 1890s, when the pit employed 179 underground workers and fourteen surface men. By the end of the First World War the colliery employed sixty-seven men underground and eleven surface workers. The King Seam, the deepest seam here at 230ft, was abandoned on 20 May 1908, and the Fulledge Thin in March 1921. The pit closed in 1935.

Opposite, below: This steel-latticework overhead bridge, constructed in 1895, crossed the main Burnley to Bacup Road connecting the Broad Clough Pit with the coal staith belonging to the Old Meadows Colliery. Although it appears to be in operation here, it was dismantled for the War Effort in 1942. The tubs were hauled by endless chain haulage, known as the 'Burnley System' or more commonly ginnies. (Harry O'Neill)

Broad Clough Colliery was one of the many Bacup drift pits which worked across the main Burnley/Bacup Road from the Old Meadows Colliery, just before Bacup town centre – the two pits were in fact connected by the surface ginnies here. Originally, Broad Clough Pit was worked by Hargreaves, Ashworth & Co., and later by George Hargreaves, and in the 1890s it employed just three men underground and one surface worker. The colliery is mentioned in a list of mines for 1945 as 'Standing'. It was reopened under the NCB in the early 1950s and new bunkers, lamp rooms, stores and compounds were erected, which also served the Grime Bridge No.3 Colliery by way of the underground Grime Bridge Lower Mountain Chain Road, almost a mile away. The pit is seen here as it was in the 1950s under the NCB. (W. Rawstron Collection)

Calder Colliery was located besides the Padiham to Clayton-Le-Moors Road, near the River Calder. Shaft sinking commenced here in December 1902, the last such operations by the firm of George Hargreaves Collieries, which later merged with Hargreaves Collieries. Problems soon arose when water-bearing strata was encountered during sinking, and the shafts were not completed until 1908. Sinking was almost completed when there was a terrible accident with the loss of four lives in February 1908. Shaft sinkers Henry Bushall, Patrick Burns, Herbert Todd and John Craggs were descending the new shaft in a small tub attached by only two chains when the tub collided with a beam and all four were thrown to their deaths. The photograph shows the shaft collar in position, ready for sinking to commence. Notice the rural aspect of what is now the busy A678 running alongside the old pit. (W. Rawstron Collection)

The Calder Colliery headgear in position ready for the installation of the cages, c.1909. By the end of the First World War, the Calder Colliery employed 200 men underground and thirty-seven surface workers, the manager at this time was James Whittaker. (W. Rawstron Collection)

A view of the Calder Colliery c.1930. An aerial ropeway took the output from Calder Pit to extensive Huncoat Pit coke ovens until around the 1920s. Later, the two pits were connected underground and the coals from Calder were raised at Huncoat Colliery and screened there. (Harry Tootle)

A 1937 scene of an early gate and face conveyor with compressed air drive in the No.6 District at Calder Colliery. Many other local pits at this time were still using chain haulage to get the coal to the pit bottom. (W. Rawstron Collection)

A similar scene at the main road loading point in the No.6 District at Calder Colliery in 1937, again showing the use of compressed air drives for the conveyor belt. These early underground scenes are rare in the Burnley Coalfield, and show the well-supported conditions of the main underground roadways. At this time the Calder Colliery was employing 318 men underground, and thirty-one others worked on the surface of the pit. (W. Rawstron Collection)

Main gate roadway at Calder Colliery showing, conveyor belts in use. Notice too the system of roof arching, in use c.1937. Conditions further inbye (the term used for the place nearer the coalface) would have been much lower in height. (W. Rawstron Collection)

A group of miners at the surface lamp room at Calder Colliery during the mid-1930s. It appears that cap lamps at this time were reserved only for pit officials, with cloth caps and oil lamps for the rest! The sign reads: 'Search your pockets for matches before going down the shaft and leave them with the banksman'. It was illegal to take any smoking materials underground due to the risk of explosions. To be caught with matches or cigarettes while being searched at the shaft top could mean that you were sent home for the day at best – or sacked at worst. Many of the miners were, however, heavy smokers, and so when underground they chose the smoking substitute, chewing tobacco, which rotted the teeth. The more 'canny' miners chewed the tobacco underground, saved it and dried it out in the lockers on the surface, and then smoked it in their pipes, or even crushed it up and used it as snuff.

Clifton Colliery pithead baths were opened on 18 February 1939 by G.C.M. Barlow, chairman and managing director of Hargreaves Collieries. The miners were instructed that:

> On the first day the baths are opened, go to work in your pit clothes, but carry your clean clothes with you in a parcel. Enter the building at the Clean Entrance. There you will receive, in exchange for your Bather's Registration Slip, a key and a card bearing the number of your two lockers. Proceed to the locker in your clean clothes locker room ... After the first day, go to work in your clean clothes, and take them off in the clean clothes room and hang them in your locker. ... A door leads out from the pit clothes room to the pit entrance. Here you will find taps for filling your water can, lavatory accommodation, boot cleaning and boot greasing apparatus.

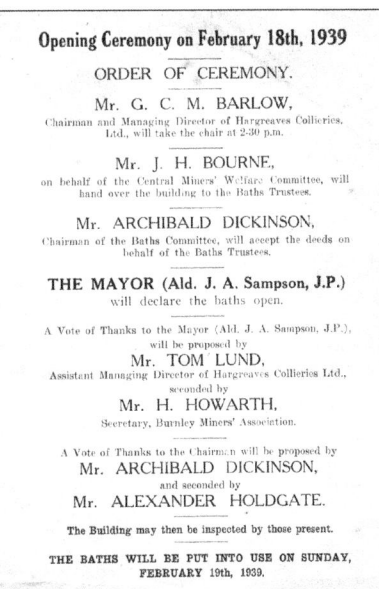

Copy Colliery, Cliviger, was described as being 'one of the NCB's oldest working units' at the time of Nationalisation. Old drift workings are known to have existed around here in the 1830s, driven into the outcrop of the Dandy Seam, most probably by Matthew Jobling, whose family were acquainted with mining at Cliviger for generations. Two shafts were sunk 96yds to the Arley Seam in 1860. Water was a constant problem in these deeper workings, and a Cornish Beam Engine was installed which proved more than adequate. One morning water broke through the shaft lining and flooded the pit, water rising 50ft up the shaft. The Cornish Engine was put to work and, after a week of working day and night, the pit was finally de-watered. Here we see the old 'back-to-back' headgear at the colliery. (W. Rawstron Collection)

Above, left: The winding arrangements at Copy were arranged 'back-to-back', that is, the engine house had one winding drum and ropes went to both shafts and cages. As the winder began to wind, one cage would be raised and the other lowered – there was one cage in each shaft. This is the old steam winder at the colliery, probably decommissioned and scrapped before the shafts were filled in, around 1958. The engine was a single horizontal cylinder engine with four valves, the steam valves on one side and the exhaust valves on the other. A rocking shaft above the cylinders operated the valves and was worked by a long handle by the engineman. To prevent the engineman becoming absentminded and letting the engine run away, there was a link motion on the engine which required regular oiling but did nothing as far as the working of the engine was concerned. In its early days the colliery had no mains water supply; water was collected from a number of springs and issues on the high moors across the road and then piped down to the pit, crossing the Lancashire & Yorkshire Railway. The water supply was also used by the nearby Copy Cottages. (W. Rawstron Collection)

Above, right: Pre-dating 1869, Cornfield Colliery, Burnley, abandoned mining in the Bing Seam on 9 February 1897. At this time the colliery, worked by the Exors of Colonel Hargreaves, employed thirty-eight men underground and thirty-two surface workers. Although mining ceased in 1897, evidence suggests that the pit was maintained as a pumping shaft, probably for nearby Habergham Colliery, until the shaft collapsed in March 1933, taking all the pumping equipment with it. The two shafts here were both 70yds deep, the upcast 10 ½ft in diameter, the downcast 12ft in diameter. The coal was raised at the downcast shaft in 3cwt tubs by a system of endless chains, which descended the pit driven by an engine with two vertical cylinders 14in by 24in and geared at one to six. This chain haulage system extended about three quarters of a mile underground. There was another smaller winding engine at the pumping shaft fitted with cages and safety appliances. It was said there were 'about one hundred men below ground' when George Henry Wilkinson was fined 20s and costs for taking the top of his lamp down in 1872. Output from the colliery was conveyed using a ginney system that ran from the pit through Grove Wood at Padiham, to Padiham Coal Staith. Many parts of this ginney embankment can still be traced in the wood, including an excellent stone-built 'turning block'. (Harry Tootle)

This delightful photograph, besides showing a group of children 'setting up camp', shows East Pit, Padiham, in the background. According to the Catalogue of Plans of Abandoned Mines, East Pit, along with Nook I' Th' Holme Pit, abandoned an unnamed seam in 1862, both being worked by the Exors of John Hargreaves. However, the pit was retained for many years after this date as a pumping shaft, possibly for the Habergham or Cornfield Colliery. The pit was one of the very early mines worked around the Simonstone area by the Shuttleworths of Gawthorpe Hall, who in fact once owned East Pit. (Duncan Armstrong)

Opposite, above: George Hargreaves & Co.'s Gambleside Colliery, located on the far side of Clowbridge Reservoir, mined from the Upper and Lower Mountain Mines. In the mid-1890s the colliery employed thirty men underground and two surface workers. By the end of the First World War thirty-three men were employed underground with eleven men on the surface. The pit also worked from a number of drifts and shafts working southwards alongside the reservoir. Here we see the last of the Gambleside Pits, abandoned in 1936. The building behind the headgear was the colliery manager's house. Rumour has it that the manager once complained to the owners of the pit that the roof of his house was letting in water – the owner gave him an umbrella. The colliery took its name from the former hamlet of Gambleside, which was mostly populated by handloom weavers and miners. A surface chain road ran from the colliery to the Swinshaw Colliery, where it continued down to the Crawshawbooth Coal Staith. The Swinshaw Colliery was disused by 1918, but was formerly also worked by George Hargreaves and mined the Lower Mountain Mine. (Harry O'Neill)

A group of Gambleside colliers. The lads, especially those in the front row, look extremely young for pit work and may have worked in pairs drawing the tubs for the colliers. The photograph probably dates from around the 1920s. (Harry O'Neill)

Habergham Colliery, Burnley, known locally as Cheapside Pit, was established by the Exors of John Hargreaves, with shaft sinkings being completed around 1873. The two shafts were sunk down to the Arley Mine at a depth of 714ft, the pumping shaft being fitted with a beam engine. A cupola shaft, fitted with an underground boiler, provided ventilation at the pit, a method used until around 1900 when a fan was installed. In later years a surface drift was driven on the site to the Top Bed Mine. The colliery was still mining in the Arley Mine in the mid-1890s, when the pit employed 318 men underground and forty-six surface men. The pit was linked underground to the Gannow Colliery, and here the coal was raised and taken either to Smallshaw Railway Sidings or the Gannow Canal Wharf. Coal could also be sent by way of an endless chain ginney track to the Padiham Coal Staith, passing the Cornfield Colliery on the way. The local newspaper announced the closure of the colliery in August 1941 – coal being extracted in later years at the Fence Drift Mines. There are no known photographs of Habergham Colliery, but this Habergham collier poses just outside the pit walls after a long, hard shift, sometime around the 1930s.

A 'posed' photograph at Hapton Valley, showing the flat winding ropes at the No.1 shaft at the pit. Notice the 'ladders' on the back stays of the wooden headgear where the men involved in maintenance to the winding wheels would have to climb. A number of 'tallies' can be seen hung up on the wall – these indicated to the management which tubs of raised coal belonged to which miner. The miner was paid for the amount of coal he raised – any dirt in the tub and he could be fined, or not paid at all for that tub. The man in the suit is probably the manager or a surface foreman. One particular tale is told about a Hapton Valley miner who was sent home for three months by the manager 'until he improved his conduct' after the miner had complained over his working place at the pit. The following day the miner, being the sporting type, set off for a bit of shooting on the slopes of Hambledon Hill behind the pit. As he made his way down the ginney road towards the pit, his gun under his arm, he noticed the manager come out of his office and quickly pin a note on the door. Curious, the miner sneaked up to read the note on the door, which said: 'OK, you can start work tonight'. Evidently, the manager thought his time had come! (Peter Nadin)

A serious fire on the surface at Hapton Valley Colliery in August 1908 caused some damage to the pithead winding apparatus. Happily there was no loss of life, although there was concern over the nineteen colliers who were underground at the time, as smoke from the fire was being drawn into the pit workings by the fan, threatening the men below with suffocation. Martin Clough, the colliery manager at the time, was away on holiday, but his counterpart at Clifton Pit, J.W. Jobling, was soon on hand. He ordered the fan to be stopped, which prevented the smoke from going down the pit but allowed gas to build up in the workings. A number of officials were sent underground by way of the ginney drift at the back of the mine, and arrangements were made for all the men to make a safe exit this way. Hapton Valley was, at this time, just outside the boundary of Burnley, and therefore the local fire brigade were not bound to attend. However, Superintendent Stedman turned his brigade out anyway when he heard that lives were in danger, but, owing to the roughness of the roads and the darkness, the fire had already been put out by the time the brigade arrived. The damage caused amounted to around £1,000, and getting things back to working order took around three weeks. (W. Rawstron Collection)

Opposite, above: Hoddlesden Collieries, near Darwen, were sunk by Joseph Place around 1838. There were no less than twelve Hoddlesden Pits altogether and they produced fireclay besides coal. Hoddlesden No.12 Pit, located on Hoddlesden Moss and known locally as 'Moss Pit', was the last of the Hoddlesden Collieries. The pit survived into Nationalisation, when it appears that another surface drift was driven. Another earlier surface drift was driven c.1933. The pits worked the Upper and Lower Mountain Mines and produced fireclay and stoneware. The pit employed over 150 men underground and twenty-odd surface workers at the turn of the twentieth century. Coal was conveyed from the Moss Pit (No.12) over Hoddlesden Moss on a chain road to the pipe works at the village. It must have been a very bleak tramp for the miners over the Moss in the depths of winter – they might (illegally) hitch a ride on the tubs, but these only travelled around 3mph. (Trevor Longworth)

Huncoat Colliery, between Burnley and Accrington, was sunk between 1890 and 1893 by Messrs Hargreaves & Ashworth, later George Hargreaves & Co. This is the earliest known photograph of Huncoat Colliery, taken c.1896, showing the old wooden headgear at the upcast shaft before it was walled in and the airlock installed. The coal-winding shaft can be seen on the left. The two shafts – one for pumping, the other for winding coal and men – were sunk to a depth of 276yd. The pit mined coal from the Upper and Lower Mountain Mines and employed 255 men underground and sixty-three surface workers. (W. Rawstron Collection)

Above: This is how the London Illustrated News depicted the disaster at Moorfield Colliery on 7 November 1883 when almost seventy men and boys died in a methane gas explosion which tore through the workings. So horrific were the injuries to some of the men and boys involved, that many of the womenfolk couldn't recognise their own husbands and sons. There was only one shaft at the colliery, and this was blocked by the cages which jammed in the blast – all the victims had to be brought out through the Whinney Hill Colliery, three quarters of a mile away. A memorial to this disaster can be seen in St James's Church at Altham, and a plaque recalling that fateful day stands close by on the roundabout. Sinking was begun at the colliery in July 1879, but no coal was raised until July 1881. The colliery was one of the first local victims of the new NCB, being closed down on 21 January 1949. The underground galleries, which witnessed those terrible events almost seventy years before, were allowed to flood. The water was then pumped to supply the coke works built on the site, the capped shaft being all that remains of the colliery today.

Opposite, above: A pre-Second World War scene at Huncoat Colliery of an underground roadway, with an early conveyor installation in the No.4 Dip workings. Dip workings occurred where seams of coal dipped away deeper into the mine – not the best method of mining coal as the water collected at the lower end of the workings where the miner worked. Wooden roof supports are still evident, steel arches not yet being in use. After sinking the shafts the water was raised from the pit bottom by tanks attached to the bottom of the cages. The winding engines were a pair of Cornish valves. The 26in-diameter cylinders had a 5ft stroke and 14ft-diameter winding drums. The upcast shaft had a Walker 'Indestructible' fan, 18ft in diameter and 7ft wide. The steam for these engines was raised by three Lancashire Boilers, 30ft long and 8ft wide, working at a pressure of 80lbs per square inch. (W. Rawstron Collection)

Opposite, below: Nabb Colliery, at Water in Rossendale, was owned by George Hargreaves & Co. as a drift mine working the Lower Mountain Mine. A chain road ran from the pithead to a staith close to Nabb Bridge, after which the colliery was named. Part of this chain road ran through tunnels, some of which are still visible, especially near the former coal staith. This photograph from around 1920 is simply titled 'Nabb Pit Boys', and some of them appear to have been very young indeed. W. Rawstron, a relation to W.E. Rawstron who preserved many of the photographs in this book, was the colliery manager at this time, the pit employing thirty-eight men underground and nine surface workers. (Steven Spencer)

Above, left: Nabb Colliery was in a very exposed hillside position, which must have been a bleak tramp for the miners going to and from work, although it was by all accounts, a 'happy pit' to work at. The picture shows three of the miners at Nabb Pit c.1943. Left to right are Billy Corless, Steven Spencer and Stanley Broxton. Around this time the colliery employed fifty-six men underground and nine surface workers. (Steven Spencer)

Above, right: More Nabb miners, left to right: John Williams, Billy Gledholme, Jimmy Cropper, Steven Spencer senior, George Howorth and Billy Place after a hard shift and in typical pit clothes – notice the bait boxes. The entrance to the chain road can be seen in the background, and the pithead baths on the left were built around the end of the Second World War. Following Nationalisation the Nabb Colliery employed fifty miners underground and seven surface workers. The manager was Stephen Landless, the undermanager J.H. Whipp. (Steven Spencer)

Opposite, above: Nabb Colliery dated from 1860 and closed in March 1954, when the manpower at the pit stood at twenty-seven. In its last full year of production the pit produced 5,124 tons of coal. The photograph shows, from left to right, Nabb miners Sam Fletcher; John Barker; -?-; Stanley Binns; Stanley Burdis; -?-. Stanley Burdis left Nabb Pit for employment at Bank Hall Colliery but, in October 1953, the thirty-four-year-old was killed following a fall from the side of the pit. Stanley had been in charge of some of the development work involved in the ongoing reorganisation at the colliery at this time, and was working on enlarging some roadways when the fall occurred. Stanley was killed almost instantly. (Steven Spencer)

Mines Rescue was an important part in any mining community and could be called upon at a minutes notice. The picture is thought to be of the local Mines Rescue Team based at Burnley, and although the equipment appears to be very primitive, it would have been the best available at the time. The Burnley Rescue Station was on Accrington Road near Smallshaw Lane, and went under the title of 'The Lancashire and Cheshire Coal Owner's Rescue Station'. This is mentioned in a trade directory for 1923, but it appears to have been disbanded soon after this date. The Rescue Station was opened in 1914 by Sir John O.S. Thursby, and the then Inspector of Coal Mines for this area, Mr Gerrard. (W. Rawstron Collection)

The ginney head underground at Old Meadows Colliery. The photograph shows the return wheel mechanism of the endless chain haulage system, and the small tubs in use at the colliery. In the 1950s this method of underground haulage was considered out of date in other local mines, as by then conveyors were being used, but it continued in use at Old Meadows right up to closure. Adaptations of this type of haulage included the endless rope system, but even this was considered obsolete on Nationalisation, and in most pits was only used for taking supplies to the coalface. In the mid-1950s the colliery was producing 15,000 tons of coal per year from the Lower Mountain Mine, and employed sixty-six underground men and twelve surface workers. (Harry O'Neill)

Opposite, above: Another scene of the underground ginney head at Old Meadows, showing the battered tubs in use at the colliery. Notice, too, the disk-shaped wheels, not the ordinary flanged wheels in use at many other of the Lancashire Pits. These ran on L-shaped rails rather than the traditional 'railway'-type rails. The tubs were turned around on steel landing plates at underground junctions. The ginney is the continuation of the surface ginney, and daylight can just be seen in the background. No doubt these men had the privilege of having their bait, or lunch, in the pit-top cabin just a short distance from the entrance to the mine. (Harry O'Neill)

Opposite, below: By the time of Nationalisation, the workings at Old Meadows Colliery extended many miles underground. The roadways were only just high enough to take the tubs – removing additional rock to make them higher cost money, something the owners of the pits were loathed to do. Walking many miles underground with back bent almost double not only tired the men out before they got to the face but took time – time in which coal could have been mined. To this end the men at Old Meadows Colliery were issued with 'trams'. By kneeling on these, and propelling themselves forward by kicking each rail sleeper, the men arrived at the face quicker and more refreshed. (Harry O'Neill)

A pit deputy 'poses' on his tram at Old Meadows Colliery. A good 'trammer' would be able to get up some speed even when going uphill. Trams in the Burnley area were fitted with flapped boards on each side, pressing these down on the wheels acted as brakes. These don't appear to have been fitted at Old Meadows – control coming down steep inclines must have been pretty well hit and miss. The L-shaped rails are evident – notice too the fine example of the old miners skill of pack walling on the right and the system of roof supports used. (Harry O'Neill)

Two miners inspect a section of old and abandoned workings at Old Meadows Colliery. Notice everything has gone – no roof supports, pipework or cables. These would all have been salvaged and reused somewhere else in the pit. Because of the system of pillar and stall mining used at Old Meadows, these workings might stand for decades before finally collapsing. The 'pillar' was a block of coal left behind to support the roof, the 'stall' was the actual area where the coal was extracted. Gas, or methane, was virtually unknown at Old Meadows Pit. (Harry O'Neill)

A magnificent example of the old miners' art of stone arching at Old Meadows Colliery – in all probability this roadway will still be standing today. These sorts of skills were lost with the passing of the old miners and the introduction of steel arches, although the latter were never used at Old Meadows Pit. This might be a section of the old drainage adit at Old Meadows, which can still be seen on the surface besides the Burnley Road. Stained a vivid reddish brown, water from the mine flowed into the River Irwell which runs through Bacup. (Harry O'Neill)

The surface ginney takes the coal from the mine at Old Meadows Colliery down to the coal staith near Meadows Mill, in the background. Notice the old haulage chain which has been utilised for a makeshift fence. It must have been very tempting for the local youngsters to ride these tubs when no one was looking, although at least one child was killed doing this when he slipped and was dragged under the tub. Screens and new bunkers were installed near the coal staith in 1960, and a new drift was driven in connection with the pit in June 1963 at Sharneyford, Bacup, giving some idea of the extent of the workings at the colliery. The days of mining here came to an end in May 1969; some of the pit's relics were taken and used as exhibitions at the former Salford Mining Museum at Buile Hill. The pit at this time had a manpower of forty-nine men, and in its last full year of production mined 15,137 tons of coal. (Harry O'Neill)

If ever any one symbol depicted an industry, then surely the most recognisable must be the colliery winding headgear. Reedley Colliery – the silhouette of its upcast shaft headgear seen here – was sunk in 1897. It was located besides the Leeds & Liverpool Canal, Burnley, and was one of the Exors of John Hargreaves collieries. This description of the mine was given in the *Colliery Guardian* on 23 September 1892:

> Two shafts are sunk here, 11ft. and 14ft. in diameter and 210 yards in depth to the Arley mine. They are situated three quarters of a mile north of the Bank Hall shafts and are downcast in connection with the latter, for ventilation proposes. ...
>
> Winding and safety appliances: The main winding engine ... has two horizontal cylinders 28in. by 60in., plain drum 15ft. in diameter, fitted with steam brake, foot brake, each brake race being 18ft. in diameter, automatic apparatus prevent over winding, and cages with Owen's catches attached similar to those at Bank Hall Colliery. The winding engine at the pump shaft has one vertical cylinder 14in. by 24in. geared 1 to 3, 5ft. drum. Safety catches are fitted to the cages.
>
> Water lifting: By means of horizontal rods and two L legs, the [pumping] engine raises water in four buckets lifted from a depth of 210 yards, diameter of each bucket being 18in. About 500,000 gallons of water are raised daily, equal to 350 gallons per minute, making a total of 580 gallons per minute raised at the two collieries. Five Lancashire boilers supply steam to these engines at 50lb. pressure.

(W. Rawstron Collection)

Rishton Colliery, Rishton, was sunk in 1883. On 29 November 1884 the *Blackburn Standard* reported that:

> On Saturday a dinner was held at the Walmsley Arms, High Street Rishton, to celebrate finding coal at the new colliery, Rishton, worked by Peter Wright Pickup. About two years ago, the old mine, which was worked by the Dunkenhalgh Colliery Co., was worked out. Fourteen months ago workmen employed by P.W. Pickup began to sink a new coal shaft, and on Thursday came across a good seam of coal.

The 'old mine' mentioned was the Meadow-head Colliery at Rishton. The new pit, at the top of Warmsley Street, was sunk to a depth of 525ft and mined the coal from the Lower Mountain Mine. 500ft below the Rishton War Memorial is a point named West One where four different haulage roads led off into the deeper workings of Rishton Colliery. The large house on the left, seen here around the turn of the nineteenth century, is the colliery manager's house, which still survives. The colliery manager would probably have been William Pickup, Peter Wright Pickup's son. At this time the colliery also mined fireclay and employed 240 underground and thirty-eight surface workers. (Gordon Hartley)

Peter Wright Pickup was the son of Peter Pickup and the junior partner in Brooks & Pickup of the Towneley Colliery at Burnley. The Rishton Colliery in the Colliery Year Book for 1938 lists the colliery directors as being P.W.P. Pickup, G.C. Pickup and G.T. Pickup. The secretary was F. Balmford; chief engineer, J. Ward; and F. Felling was the colliery manager. Peter Wright Pickup is also listed as being the manager at the Meadow-head Colliery at Rishton in 1882, abandoned in August 1884 when the new pit at Rishton came into operation. (Gordon Hartley)

Rishton Colliery, c.1910. A more leisurely age, with horse-drawn carts taking the coals raised at the pit to the local mills and household customers. The colliery was closed down due to uneconomic workings in 1941. Water was pumped from the old Rishton mine shafts and used at the nearby Rishton Paper Works from around 1970 to 1983, when the latter closed. The pit shafts were eventually filled in in August 1991 to make way for the construction of an access road the South Side Housing Estate – severing forever Rishton's links with the coalmining era. On 14 September 1955, the *Northern Daily Telegraph* reported that:

> *A new road has been made opposite Phillips Road, Blackburn. The shaft at the Blow Up Pit is covered in, about one yard from the new road, and about 60 yards from the arterial road. The water in the pit is standing about thirty-two feet from the surface. The old colliery workings from Sett End, Whitebirk, and Blow Up are full of water, and water is still being pumped at Rishton Colliery to Dean Reservoir at Great Harwood, for the Accrington and District Water Board.*

The Blow Up Pit, actually the Little Harwood Colliery, was so-called by the locals after a boiler burst in 1819, killing John Landless, Thomas Pilling, John Tithrow and William Wood. There is a memorial to this sad event in the Parish Church of St Peter's at Burnley. (Gordon Hartley)

The only know image of Burnley's Rowley Colliery, located opposite Rowley Hall, Burnley, and another of the Exors of John Hargreaves Pits.

The colliery was sunk between 1861 and 1862 to the Dandy Mine at a depth of 310ft, the last tub of coal being raised on 8 May 1928 following the exhaustion of economic workings. Just after the First World War the colliery employed sixty-three men underground and thirty-two surface workers. Following closure, the pit remained dormant for many years before being demolished, and the site was used as a tip for the colliery spoil during, and for many years after, the reorganisation at Bank Hall in the early 1950s. The author remembers, as a lad in the late 1950s, going up this tip to collect bits of coal, tipped among the debris, in an old trolley made out of wood and pram wheels, to earn a bit of pocket money. We sold the coal to our neighbours for 5s (25p) a bag, and on a good day we could earn maybe a pound between us. Many adults were also there, picking out bits of coal to fuel their homes from the muck and stones of the tip. I think this was illegal, as well as dangerous, but no one bothered us. Each time a load from the NCB wagons made a tip, scores of people rushed forward to try and get the best pickings – even as the wagon was still tipping its load. On one occasion, after hours of work picking coal, the wheels of our trolley collapsed as we made our way down the steep incline which led up to the tip. Those on the way up the tip, seeing our predicament, stopped to have a cigarette, or a pretend 'rest', biding their time until we left and they could collect our spilt 'booty'. Bedraggled, tired and somewhat angry, we eventually had to leave the scene – and our precious sacks of coal! (Joan Barber)

Taylor's Green Colliery was located in an area with a long tradition in coalmining, to the north-west of Blacksnape, above Darwen. Dating from the 1870s, the pit mined the coal in the Half Yard, Little Ten Inch and Lower Foot coal seams. The pit was worked under the title of Thomas Knowles' Spring Vale Fireclay Works and, although predominantly a fireclay mine, coal was worked up till 1943. Only fireclay was extracted after this date until abandonment in the early 1950s. Just fifteen men worked underground in 1945, with two surface men. The pit is last mentioned in the *Guide to the Coalfields* in 1948, when it employed twenty men. Thomas Knowles' firm continued to make various clay goods after the pit had been abandoned, until 1970. There were at least four shafts used at the Taylor's Green Colliery, which is today covered by a housing estate.

Top O' Th' Coal Pits was an old coal pit located on what was then named Coal Pit Moor, between Darwen and Blackburn. This sketch, the earliest known, depicting coalmining in East Lancashire, is a rudimentary image of Top O' Th' Coal Pits by Charles Haworth, dated January 1846 when the pit was abandoned. The 1844-1848 OS map shows no less than six 'old coal pits' around the area, all in close proximity. The horse gin used to raise the coal is obvious, although abandoned at this time. The 50ft-deep shafts were rediscovered when Blackburn Corporation built Park Lee Hospital on the site. The seams varied from 2ft 3in to 4ft high, while the underground galleries ran a distance of about 50ft towards and under the new hospital grounds.

Towneley Colliery was established by Brooks & Pickup on 25 February 1869, when the first sod of the 'Alice Pit' was cut by Miss Towneley, later Lady Alice O'Hagan of Towneley Hall. Evidence suggests that Brooks & Pickup had already begun mining in this area at the nearby Boggart Bridge Colliery, having negotiated with the Towneley's for the right to mine coal there. Around the late 1900s the pit employed 496 men underground and a further 148 on the surface. Here, in a school outing to the colliery, the children look as if they are going on an underground trip – the flame safety lamps are very evident for the descent into the workings. There were three shafts at the colliery: the 'Alice Pit', the 'King and Dandy Pit' and the pumping shaft. 'Alice Pit' was the deepest, and was upcast in ventilation. The fan house was located on the surface near the mouth of the shaft. At the pumping shaft was a two-deck cage capable of carrying sixteen men, eight on each of the decks. Wooden headgearing was still common at this time, but steel was taking over and in fact was used at both the 'Alice Pit' and the pumping shaft. However, the 'King and Dandy Mine Shaft', or No.3 Pit, had wooden headgear. The 'Alice Pit', or No.1 Pit, was 820ft deep and 12ft in diameter. The Fulledge Thin Seam (197ft), Lower Mountain Mine (379ft), Dandy Mine (493ft) and Arley Mine (770ft) Seams were all intercepted at the 'Alice Pit.' The 'extra' 50ft in depth was utilised as a sump, or water catchment. The No.2 Pit, or pumping shaft, was 825ft deep, the sump being 73ft below the bottom of the Arley Mine. Water from the 'Alice Pit' was pumped over to the No.2 shaft sump, and thence pumped to the surface. The 'King and Dandy' shaft was also 12ft in diameter, but just 497ft down to the Dandy Mine, below which another sump was sunk to give a total depth of 554ft. At the pumping shaft was a Cornish Beam Engine, the rocking arm weighing about 15 tons. One end of the arm was coupled to the flywheel and piston that operated in a cylinder with a 6ft stroke. The other end was coupled to the vertical beam suspended in the shaft, and this moved the ram up and down in the suction pipes. (Burnley Borough Council, Towneley Hall Museum)

Townley Colliery had extensive sidings besides the pit's surface ginney, seen here around the turn of the twentieth century, whereby the coal was sent on by rail on the Lancashire & Yorkshire Railway. Messrs Brooks & Pickup probably had some regrets on sinking the Towneley Colliery for, in 1873, five years after the first sod was cut at the pit, the miners went on strike in what was to become one the most bitter disputes in Burnley's coalmining industry. During this strike, the result of a mixture of Union recognition and the right to have the colliers' coal by weight or measure, hundreds of local miners went on strike. The colliery owners retaliated by bringing in scab labour, or 'knobsticks', from as far away as Devon and Cornwall. In fairness, the import labour knew little or nothing of the strike until they arrived in Burnley, and, while a number were persuaded to go home, a number stayed on, facing intimidation and threats of violence and death. The 'knobsticks' were housed in various parts of town, including Cornwall Terrace, also in an area which soon became known as Little Cornwall, off Rossendale Road, and in cottages on the Ridge and at Cliviger, in houses known as Lower Damfield. (Gordon Hartley)

Opposite, above: Along with Copy and Railway Collieries, the Union Pit was associated with generations of the Jobling family. This image is believed to show John Jobling, the pit manager in 1896, during the 1921 Coal Strike. The pit at this time employed seventy-nine men underground and eighteen surface workers. In 1866 three people – Joseph Ion, Thomas Alderson and James Carlisle – were suffocated at the Union Colliery when a fire started at an underground furnace. I am at a loss as to how the system of ventilation was achieved at the colliery; the shaft shown here appears to be downcast and the drift over by the Railway Colliery was probably also downcast. It might be, therefore, that the Hole House Pumping Station just up the road was also used as an upcast shaft fitted either with a furnace or a fan. (W. Rawstron Collection)

Opposite, below: A more general view, taken around the 1930s, of the Union Colliery and its surface arrangements. The pit headgear now has two winding wheels, and therefore two cages. Over to the left can be seen the cupola chimney, which provided the ventilation for the Cliviger Coal Co.'s Railway Colliery. The chimney collapsed one day without warning, however the ruins can still be seen today. (Peter Nadin)

Three
Nationalisation: Out with the Old, In with the New

The coal mines of Great Britain were Nationalised on 1 January 1947 to be 'Run by the Government on behalf of the people', as the new National Coal Board proudly announced. Many miners and management saw this move as the industry's 'New Jerusalem', which never did appear, but at least some investment and improvements were made as far as the miners were concerned. The country at this time was on its economic knees: the war had only finished a few years before, and production at mills, foundries and factories had been run down to the bare essentials as far as output went, as the men left to fight the war, the return to full production being a slow, painful process. It was coal once again that was going to put the 'Great' back into Britain – and every ounce of it was needed. The cry went out 'Coal, more coal', and, to get the country back on its feet, the miners responded immediately by working six, even seven-day weeks, only to be betrayed by successive governments and the rundown of the industry in later years. A number of local collieries were closed down soon after Nationalisation through underinvestment and being outdated, and new drift mines were opened-out around the Burnley Coalfield. Those pits considered to have a descent lifespan were reorganised, and heavy investment brought them up to date. This advertisement, dating from just before the pits were Nationalised, asks people to burn local coal from Lancashire pits.

Burn Coal

BUT

see that it is Lancashire Coal from the Local Pits

Be Patriotic and Help Local Industry. You will be helping your friends or even yourself.

LOCAL COAL HAS NO RIVAL.

Lancashire Associated Collieries,
SALES OFFICE - - HUNCOAT, near ACCRINGTON.
Telephone—Accrington 3291.

1

Vesting Day, as the day on which pits were Nationalised became known, was marked at Bank Hall Colliery with a ceremony on Sunday 5 January 1947. Left to right are: W.E. Rawstron, colliery manager; W. Grant for the National Union of Miners (NUM); T. Whiteley, checkweighman; and the youngest underground worker at Bank Hall, A. Wright. (W. Rawstron Collection)

Among the first of the Burnley area collieries to be modernised was Bank Hall Colliery, seen here in September 1948. The outdated Arley screens and landsale depot can be seen, a constant source of dust complaints to the management of the pit from the residents at Browhead. When the Council threatened the then owners, Hargreaves Collieries, with court action, Hargreaves replied that the cost of new screens was prohibitive, but that they could, of course, always close down the pit. The court case was subsequently dropped. Notice the piles of chocks and other timber roof supports, and what seems now to an ancient NCB wagon. The No.4 shaft at Bank Hall can be seen in the background. (W. Rawstron Collection)

The planned project for reorganising Bank Hall Colliery was going to be a massive undertaking, costing over £1 million. This was to include a complete new pit bottom at the No.4 shaft to cope with the new 5-ton mine cars, along with reorganisation of the surface layout to handle these new larger tubs. New screens and washers would also have to be built – all without interrupting the general day-to-day running of the colliery. The men were raised and lowered down the No.1 (upcast) shaft at the pit, shown here, while this work was going on. The shaft, enclosed in brickwork, stopped the fan installed in the white building from drawing air down the pit from the top of the shaft. The sloping structure besides the building, a fan drift, connected with this shaft a few yards below the surface and extracted the air through the workings. (Burnley Borough Council, Towneley Art Gallery and Museum)

The old and the new washers, seen here in March 1954, stood side-by-side for a number of years. In the old screens the coal was sorted by hand. The new washer is in use here, as can be seen from the fleet of NCB wagons waiting to be filled with clean coal. At this time the coal preparation plant, as it was to be named, was handling 30,000 tons of coal a month, 7,000 tons of this being discarded as waste normally passed on to the customer. The plant handled the coals from Bank Hall, Union Mine, Reedley and Dandy Mine. The coals were 'processed' by gravity, electric magnet vibration, compressed air, water baths and water sprays. To stop pollution in the nearby River Brun, the water from the washed coal was allowed to settle in huge settlement tanks before being discharged into the river. (W. Rawstron Collection)

New entrance gates were constructed at the Queens Park Road end of the colliery in 1954 for the wagons using the new coal preparation plant. The road rollers still appear to be at work at this time, laying out the new road. Waste was disposed of by wagons taking the spoil along Netherwood Road, crossing the River Brun by a ford, and making the perilous journey up the steep side of Rowley Tip. The washed coal was taken either by wagons to the customers or, for more distant distribution, along the colliery rail siding to the sidings at Central Station. A small signal box on the colliery line controlled the rail traffic onto the main line. Today, the line of this old mineral railway is a pleasant footpath running from Burnley Central Railway Station to Worsthorne Moor. (W. Rawstron Collection)

Preliminary work was begun around the No.4 shaft top at Bank Hall Colliery during the annual holiday in July 1954. This was to include laying out of new concrete foundations for the mine cars circuit, and the surface traverser and tipplers. Work was still ongoing underground, and it was a race against time to get this work completed before the men arrived back off holiday. Among the other jobs in hand were the shortening of the winding ropes and guide ropes. The old cages would have to be taken out and new ones for the larger mine cars installed. (W. Rawstron Collection)

A complete new pit bottom was built at Bank Hall, 90ft above the existing one, the brick walls here being 5ft thick and the concrete floor 3ft thick. The underground tunnels at the No.4 shaft were enlarged to 20ft wide and 14ft high at one point. Here is the busy scene at the pit bottom looking towards the shaft, in July 1954. The new rails were for the larger mine cars and the locomotives. (W. Rawstron Collection)

Nearer the shaft here on the empties side in July 1954, showing the new underground lighting at pit bottom and the new mine cars in use at Bank Hall Colliery. Compared with the old shaft bottom this would be more like a modern factory than a coal mine. Most of the colliers at Bank Hall wouldn't have seen this work going on until they arrived back off their holidays. They were sent to the coalface by going underground at the No.1 shaft, and an old tunnel was brought back into use to raise the coal at this shaft while the work was ongoing at the top and bottom of the No.4 shaft. (W. Rawstron Collection)

Another view of the Colne Road entrance gates to Bank Hall Colliery, showing the on-going activity in 1955. Being built are the colliery stores, electric workshops and the new baths extension. The gates on the left-hand side still survive, and it is here that the more observant will be able to find the NCB sign, albeit covered over now with various other signs for the firms which occupy the site today. (W. Rawstron Collection)

Underground at Bank Hall there was further investment and progress with the installation of the most modern equipment in coalmining at this time. A 64hp-battery locomotive equipped with radio telephone, first used in the Towneley Tunnel in the mid-1950s, is seen here. (W. Rawstron Collection)

At Bank Hall Colliery in March 1959, four teams of six men set a new British and European tunnelling record by driving more than 66yds in seven days. The 1,400yd-long Towneley Rise Tunnel was fully opened up in January 1960. It enabled the miners to work a new area of virgin coal previously separated from existing works by a 25yd fault. The tunnelling team was somewhat of an international affair and included Tom Layfield who had spent forty-nine years in the Burnley pits, P.M. Connell and M.J. Quinn from Ireland, K. Tander and E. Priim, both from Estonia, and two polish men. The previous British record had been held by the men at Agecroft Colliery, near Manchester, and stood at 60yds advance in five days, 5yds shorter than the European record. The Bank Hall miners in fact broke the record by just 1yd. Here is the Distinction 50B Loader and Mine Car in an 18ft by 10ft section of roadway in the Rise Tunnel where the record drivage was made. (W. Rawstron Collection)

Opposite, above: The NCB was keen to show what great improvements they were making at local pits. Here at Bank Hall Colliery they displayed a selection of tubs used in the local mines. Front row, left to right, is a tub used at Deerplay Colliery, one $3\frac{1}{2}$cwt-capacity tub from Grimebridge, 4cwt tubs used at both Hapton Valley and Scaitcliffe Collieries, and a 3cwt tub from Hapton Valley. In the middle row are 18cwt tubs used at Hapton Valley and Huncoat, one 10cwt tub from Bank Hall, and another from Hapton Valley of 6cwt capacity. Back row shows the NCB's pride and joy: the 5-ton mine car. (W. Rawstron Collection)

All new recruits to the coalmining industry had to undergo sixteen weeks training, carried out at Bank Hall Colliery and Burnley Municipal College's Mining Department. Classes consisted of first aid courses and the theoretic and practical side of mining. In the case of the latter, new recruits weren't allowed to go underground unsupervised, so many were sent to other local pits to work on the surface loading props to go underground and carrying out other trivial tasks. There was, however, a training face down Bank Hall's No.1 shaft where supervised practical training was carried out. Part of this underground training consisted of pushing a laden tub to the top of an inclined road and emptying it. The trainees would then refill the tub, which would then be taken back down the incline – and again emptied. The process was then repeated. Here we see the new training centre at Bank Hall, opened in June 1953. (W. Rawstron Collection)

Training certificates were handed out at the new training centre the same month as it opened, June 1953. Here, H.E. Clegg, area agent for the North West Division of the NCB, T.I. Jeremiah, training officer and W.E. Rawstron, manager at Bank Hall Colliery, hand out certificates to the newly trained coalmining recruits J. Shuttleworth, T. Cottom, Tommy Cummings, R. Stephenson, watched by P. Littler. Yes, Tommy Cummings is the famed Burnley Football Club player – and he was also a coal miner. (W. Rawstron Collection)

Opposite, above: In May 1948, ten European miners, including Poles and Ukrainians, reported for duty and training at Bank Hall Colliery. They were housed at the new Huncoat hostel, which would eventually accommodate between 100 and 150 men brought into the country to ease the acute labour shortage in the pits of East Lancashire at the time. Ivor Jeremiah (Group Training Centre manager) is seen here showing the new arrivals around the colliery, on his right is Ernest Lord, training officer at the pit and on the extreme left is P. Littler, another training official. To overcome the language barrier, the new volunteers were given lessons in English before leaving the resettlement camps and each also received a vocabulary book printed in three languages. To avoid being sent back to their country of origin, one condition imposed on the new workers was that they had to stay in the coal industry for a certain number of years. Many, however, did stay on at the Burnley pits and were excellent workers. They were often assigned the hardest of colliery work: that of driving new tunnels and new developments underground. However, it wasn't only men who came to Britain, the following year six Czechoslovakian women came to Accrington to work in the Prospect Mill there. They were housed in the old Petre Arms pub, which had been hastily converted into a hostel. (W. Rawstron Collection)

Bank Hall Colliery looking towards the No.4 shaft during reorganisation in 1954. This would have been a daily sight to those miners employed at the colliery, who would leave the lamp room and make their way towards this shaft for the descent into the pit. Notice that the blue and black NCB flag is flying. Collieries were set output targets, and if they broke this target, even by a ton, they were allowed to fly the flag atop the colliery headgear. Bank Hall was often seen flying the flag. The 'Cathead' – the structure on top of the headgear – was erected in 1954, its purpose being to examine the winding wheels or replace them. (W. Rawstron Collection)

Among other improvements made at Bank Hall Colliery during reorganisation was the commissioning of the new gas-fired boilers in April 1957. The boilers were fired by methane extracted from the pit and pumped to the surface. Previously, this gas was just vented into the atmosphere, until it was noticed that birds, particularly pigeons, were dropping out of the sky – literally gassed. The firing of the boilers with the extracted methane saved the burning of about one hundred tons of coal per week, and a saving for the colliery of some £17,000 per year. This supply of auxiliary power was not a new thing to the colliery. Previous to and during the early 1930s, the exhaust steam from the boilers was used to drive a generator which fed the power lines used by the town's electric trams, until the buses arrived in 1934/35. (W. Rawstron Collection)

Opposite, above: Work was carried out at Calder Colliery by the NCB during this 1950s boom time. Here, shown at Easter 1957, we see the new thirty-five-ton skip-winding installed at the colliery. It was a rather belated and unnecessary expense, for in July 1958 the pit was closed down for economic reasons. In skip-winding, the cage was detached from the ropes and a skip in the form of a vertical bunker was put in its place. The skip was open at the top and provided with an automatic door at the bottom. It was filled at the shaft bottom by going deeper than normal into the sump area and fed by a shute from a storage bunker underground. At the surface the coal was discharged onto a trough-shaped conveyor belt and carried up a ramp to be screened. The screens and the belt can be seen in the middle background. (W. Rawstron Collection)

Another view of the skip-winding arrangement at Calder Colliery – as it was just after the colliery was closed down. The skip installation should have saved a great deal of time by replacing the winding of coal up the shaft in tubs, and consequently a saving in costs, however this came too late for Calder Pit. On closure the pit was salvaged of all reusable material and the men transferred to other local pits. Today, all that remains of the former Calder Colliery is the red-bricked winding house now used by small industries. (W. Rawstron Collection)

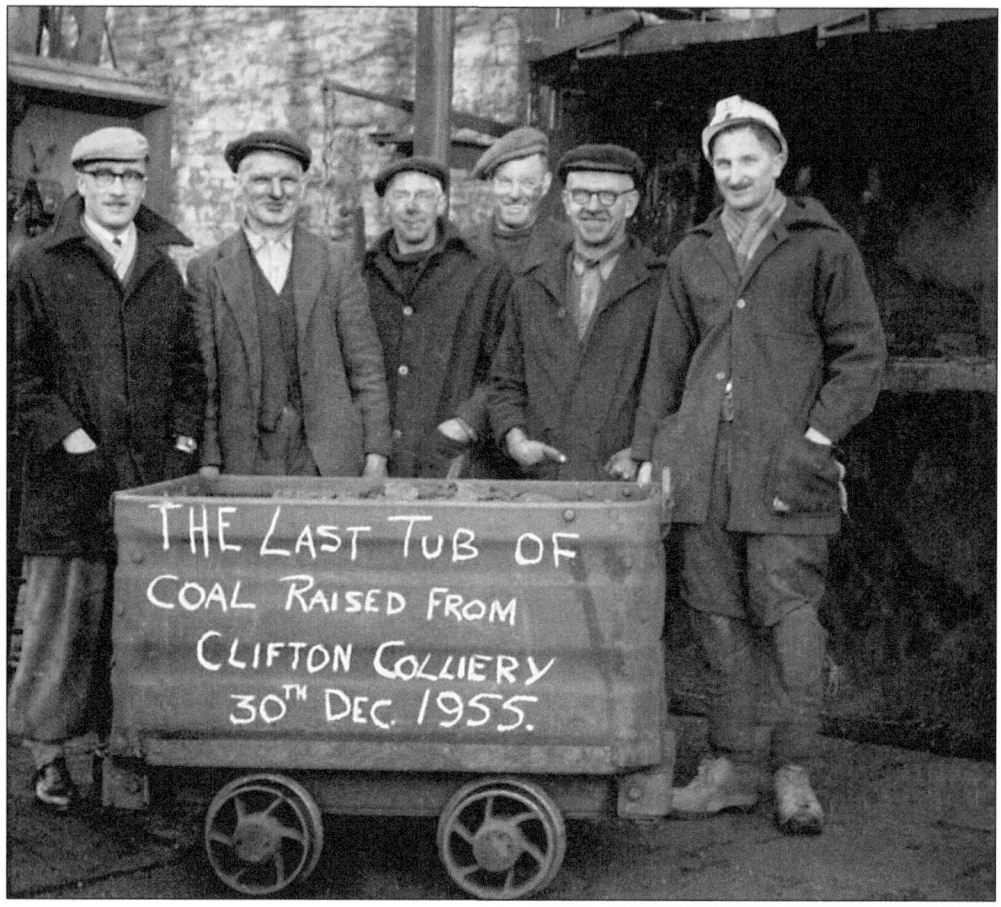

Clifton Colliery, Burnley, faired even worse than Calder Colliery, for it was closed down on 30 December 1955. Here we see the last tub of coal ever raised at Clifton and, left to right: H. Tomlinson, colliery manager; T. Boardley, banksman; A. Robinson, surface foreman; H. Fort, storekeeper, and H.F.S. Smith, colliery overman. Clifton Colliery was regarded as being a 'family pit': son followed father, and his son followed him. This tradition came to an end with the many closures, as colliers from other pits were transferred to other local pits as theirs closed down. The colliery had a number of surface ginnies, one ran to the nearby paper mill, another to the Oswald Street gasworks, and another to a coal staith located on Junction Street. (W. Rawstron Collection)

Opposite, below: A new fan drift was driven at Copy Colliery in 1948, and brand-new pithead baths, seen here on a bleak, snow-covered winter's day, were opened in February 1953. This previously undreamed-of luxury meant that the miners could go home warm and dry instead of wet and dirty, and must have been a very welcome addition at the pit. The building, which also housed the pit canteen and the first aid room, was opened by H. Howarth, the area agent for the local NUM. Presiding was colliery manager S. Jobling and also present was R. Lowe, area manager. The following Saturday the celebrations continued as presentations were made at the Gordon Lennox Hotel to long-serving miners from the pit. Among those present to receive their long-service medals and a Westminster chime clock were Orlando Crabtree, who had worked at the pit for fifty-one years, and John James Kershaw, who had completed fifty years mining service. (W. Rawstron Collection)

A surface drift was driven at Copy Colliery in 1937. Here the open cut for the drift enlargement and portal is under construction in June 1948, soon after Nationalisation. The $3\frac{1}{2}$ cwt tubs can just be made out, coming out of the drift mouth being drawn by endless chain. Copy Colliery was never a big pit, at the time this work was ongoing it employed just fifty-two men underground and seventeen surface men. The manager at the pit, which mined the coal from the Dandy Mine, was S. Jobling and the undermanager W. Collinge. Both these families were connected with mining at Cliviger for generations. (W. Rawstron Collection)

Deerplay Colliery was surely Lancashire's highest pit, located on the moors between Burnley and Bacup. The colliery was originally established by the Maden family, who sold it to Brooks & Pickup, of the Towneley Collieries, around 1860. In the mid-1890s the pit, which mined coal from the Mountain Mine, was worked by the Deerplay Colliery Co. and employed ten men underground and three surface men. In 1945, as things improved, the Deerplay Colliery employed nearly forty men. Following Nationalisation, investments were made at the pit. The pithead baths, seen here, were opened on 25 August 1956, at a cost of £17,000. A temporary structure was considered by the NCB but the colliery was considered to have at least twenty-five years of coal left to mine, so a permanent building was erected. The showers could accommodate up to 144 men.

Opposite, below: The Fence No.1 Drift was one of six drift mines planned by the NCB, which included Hill Top (Bacup); Overtown (the Salterford Pits at Cliviger); the Fence Nos 1 and 2; the Fir Trees Drift; and a further one at Wigan. The Fence No.1 Pit connected with the workings at Wood End Colliery near the Duck Pits Sewerage Works. During the driving of the pit in September 1948, when this photograph was taken, the men were down 180yds. The pit banksman, Thomas Cross, is turning the wheel on the Bunker to release the coal into an NCB wagon. Fence No.1 mined the Arley Seam, exposed as outcrop at this location. The seam was also worked from the abandoned Habergham Colliery, but the long, underground haulage system here increased costs and consequently it was decided to extract it from these drift mines. Such was the demand for coal at the end of the war, that vast amounts of coal were also mined by opencast methods – notably at Higham, Cornfield, Brunshaw and around Grove Woods at Padiham. The Fence No.1 was worked out by November 1950, and men and materials were moved to the nearby Fence No.2 Drift. (W. Rawstron Collection)

Other improvements at Deerplay Colliery included the erection of the coal bunkers and gantry in 1956, and the tarmacked colliery yard the following year. Notice the type of NCB wagons in use at this time. On more than one occasion the colliers arriving at Deerplay Pit, would, owing to the high altitude, find the whole area covered in snow and they literally had to dig their way in to work. Deerplay never did live up to the life expectancy of 'twenty-five years coal left to mine'; the pit, which at the time employed 212 men, was closed in April 1968. In its last full year of production, Deerplay raised almost 94,000 tons. It was a sad day when the Deerplay colliers had a farewell drink at the Deerplay Arms – a day that also ended over one hundred years of mining on the bleak moors. (W. Rawstron Collection)

Above, left: The Fence No.2 Drift, a few fields away from the No.1 Drift, was started in 1947. In that year a great deal of difficulty was experienced during drivage, due to glacial drift and the extremely inclement weather. The Fence No.2, which linked up with the Wood End Colliery, often produced in excess of 900 tons per week using the three-shift system. The afternoon shift cut the coal, moved over the conveyor and put the packs on. The night shift fired the shots and withdrew the roof supports. The day shift loaded the coal onto the belts. (W. Rawstron Collection)

Above, right: The Fence No.2 Drift, showing the portal of the new mine and conveyor gantry. Unlike the No.1 Pit at Fence, the No.2 used the top belt of the conveyor, as is more traditional. The heavy boulder clay and running sand caused tremendous problems during driving of the drift entrance. In the end, a dragline was brought in, and the start of the cut was close piled and concreted. The drift then dipped down into the workings at one in four to reach the Arley Mine. These drift mines were relatively inexpensive to develop at a time when the country was crying out for coal, but the schemes only had short lives – maybe three or four years before being worked out. (W. Rawstron Collection)

Opposite, above: The coal bunkers, on the left, and the dirt bunkers on the right, were erected in 1949 at Fence No.2 Drift. Notice the makeshift brazier made out of an old oil drum to keep the surface workers warm – there were no other buildings at the Fence No.2. Coal was taken from the pit to Bank Hall Colliery in NCB wagons for screening. Lights had also been put up at this time to ensure continual production, even throughout the night. All that remains of this project today is the concrete base of the coal bunker and the stubs of the steelwork legs of the bunker, where they were cut off with oxyacetylene equipment. (W. Rawstron Collection)

Left: The drift at Fir Trees Colliery dipped down into the workings at one in one and a half – a drop of 1ft in every 1½ft! The 400 sleepers acted as steps for the miners going in and coming out of the drift. The degree of incline was allegedly equal to the steepest side of Pendle Hill, or as steep as the stairs in an ordinary house. A manrider was eventually installed, which gave better access for the men and materials. Although the colliery worked the Arley Mine, out of range of the old Habergham (Cheapside) Colliery, and was dewatered by a submersible pump in the shaft at Habergham, much of this valuable coal had to be left unworked. When the old Habergham Colliery was abandoned, the pit shaft became a local dumping spot. The submersible pumps could only be lowered to a certain point in the shaft and a vast area of coal below that point was flooded forever. Around 160,000 tons of best Arley coal had been mined at Fir Trees Colliery before it was abandoned on 11 March 1966. In 1963 the colliery employed seventy-two men underground and six surface workers. (W. Rawstron Collection)

Work was begun in earnest on reorganisation at Hapton Valley Colliery soon after Nationalisation, including a new surface layout with new screens and a covered tub circuit around the coal-winding shaft. Notice that the cathead, the device above the winding wheels, had not yet been installed at the new headgear at the coal-winding shaft. This replaced the old steel lattice-framed headgear installed when the new pit was opened around 1910. The upcast shaft was enclosed with brickwork creating a new airlock, seen here in the early 1950s. The NCB saw the colliery as a long-life pit with huge reserves of coal, which warranted large investment. Under the old owners, Hargreaves Collieries, little investment was put into the pit, which nevertheless produced more coal than other local pits. All that was to change. The pit at this time was working the Old West Side and Old Rise districts. The chain haulage on the Old West Side were said to have been some 4,500-5,000yds long, and rope haulage was only introduced on the main brows during the two wars. The first longwall coal cutter was introduced at Hapton Valley in 1910, soon after the 'new' pit was opened out. It wasn't successful and was withdrawn, and it wasn't until around the 1926 coal strike that longwall machines became more common at the pit. (W. Rawstron Collection)

Opposite, below: New coal and dirt bunkers were installed during the surface reorganisation at Hapton Valley Colliery during the early 1950s. Hapton Valley never had any direct links with either rail or canal and coal at this time was still being sent, via the ginney system, from the pit to Smallshaw Railway Sidings. The new coal and dirt bunkers meant that coals could now be removed from the pit in NCB wagons. Dirt was tipped and compacted in Spa Clough to the west of the colliery, water from the stream here being piped through 6ft concrete pipes. Although the screens were said to be 'new', the dirt was still handpicked off moving belts. The author remembers being put to work on the belt as a trainee miner at Hapton Valley around 1963. Lumps of dirt were handpicked off the belt and thrown onto another belt which went to the dirt bunker, while the coal carried on and went into a crusher. Letting a coal cutter pick go past into the crusher was fun as the sparks flew and bits of metal were hurled in every direction. (W. Rawstron Collection)

In July 1954, a brand-new Metro-Vick 220hp electric winder was installed at the coal-winding shaft at Hapton Valley Colliery. In order not to interfere with coal production, the work involved in installing this engine was carried out over a weekend. Remarkably this work, which would have included the installation of the winding engine, the changing of the winding ropes, and the coupling and uncoupling of the cages, was all carried out on time. (W. Rawstron Collection)

The new tub circuit at Hapton Valley Colliery in the 1950s. As each cage arrived at the surface, empty tubs were pushed forward by hydraulic rams pushing the full tubs out of the cage. These then went forward into the circuit to a tippler to be emptied, before returning once again on the empties side of the shaft to repeat the process. Prior to this, the operation was carried out by hand, with its obvious dangers at the pithead. The winder had no direct view of what was going on at the shaft top – he relied entirely on signals from the banksman, the person in charge at the top of the shaft, and the onsetter, the person in charge at the bottom of the shaft. (W. Rawstron Collection)

Opposite, above: Other work on the surface at Hapton Valley Colliery involved felling the chimney at the old Spa Pit on 13 October 1956. Here colliery manager Harry Warne sets the charges at the chimney of the old pit, twenty-six years after it last raised any coal. On his left, undermanager Jimmy Cregg prepares the primers for the shot. Almost all the other buildings at the old Spa Pit were also demolished at this time, the rubble going to provide the new access road at Deerplay Colliery. (W. Rawstron Collection)

There she goes, the end of an old timer – but not without some resistance. In spite of all Harry Warne's knowledge of shotfiring, the old Spa Pit chimney proved to be a stubborn object to remove. Even after using 14lbs of explosives in the initial blast, the chimney stood firm, held by a 3in-wide, 1in-thick steel band. A second blast did the trick. The old wooden structure of the headgear at the old Spa Pit can be seen to the left of the falling chimney. (W. Rawstron Collection)

As the Union Mine was developed at Hapton Valley, an argument was put forward for a new surface drift – not least because the Union Mine was very gassy. A new drift would improve the pit's ventilation, and a conveyor belt would reduce the cost of getting the coal to the surface. The NCB approved the plan, which would involve an incline tunnel from the surface dipping down to meet the existing workings in the Union Mine close to the bottom of the East Side Drifts. Work was begun at the bottom end of the Municipal Cemetery close by the old Spa Pit, and here, on 29 March 1960, the first of the arches were set for the portal of the new drift at Hapton Valley Colliery. (W. Rawstron Collection)

Opposite, above: The surface drift at Hapton Valley was driven by drilling the rock face and firing it with explosives. The dirt was then loaded onto the temporary dirt belt using an Eimco bucket loader, seen here. The drift was 16ft 9in wide and 11ft 6in high, the total length being 1,260yds. Midway down the drift a point called the 'passing place' was widened to allow the manriders to pass each other. The manrider worked on a three-track system, whereby the middle track was split at this point to enable the passage of the two trains. In November 1961, while driving this drift, a (then) European record advance rate for a dipping tunnel was achieved when three teams of four men, working on a three-shift system, drove the tunnel 45yds in five days. The cost per yard in driving the surface drift, estimated at £90, was put at £65, a saving attributed to the hard work of the men and the use of the Eimco bucket loader. (W. Rawstron Collection)

Opposite, below: In January 1962 the new drift 'holed' into the new workings. Following this there was an immediate alteration in the pressure differences in all working parts of the mine. It was considered that the booster fan, located at the bottom of the East Side Drifts, was no longer required and would be taken out of use on a care and maintenance basis. In order to find out the effects on pressure and ventilation, experiments were carried out running both the main fan and the booster fan together and individually. Here the record-breaking drift drivage team are, front row, from left: Billy Lord; B. Bond; Eddie Catlow; Kenny Royal; J.B. Dearden. Back row, from left: J. Hutchinson; R. Gregory; W.H. Moor; Adam Weir (manager); Dick Bernard; N. Green; J.R. Oldham. (W. Rawstron Collection)

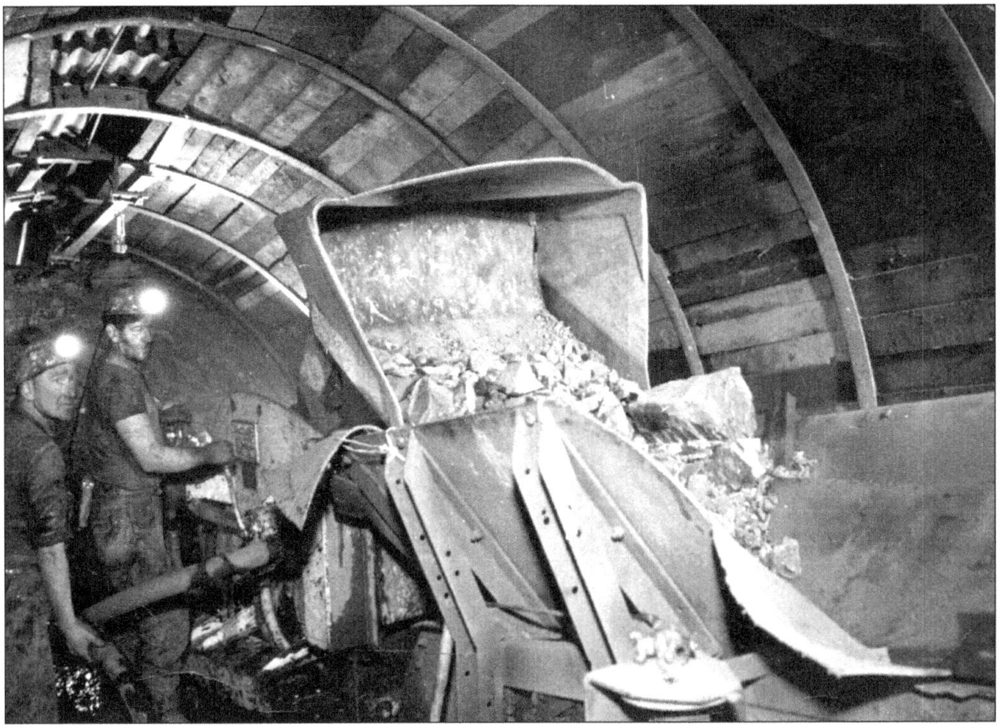

A general view of Hapton Valley Colliery on 19 January 1962 showing the surface layout of the pit. The new drift at the colliery had just broken through into the main workings of the pit. The cooling towers in the background are those of Huncoat Power Station, whose main customer was Huncoat Colliery. Most of Hapton Valley's coal was sent to the Padiham Power Station, and the Ribblesdale Cement Works at Clitheroe. The colliery at this time employed 369 men underground and sixty-seven surface workers. (W. Rawstron Collection)

Opposite, above: At 9.43 a.m. on Thursday 22 March 1962, a methane gas explosion ripped through the workings of the Rise Two District at Hapton Valley Colliery. The blast killed sixteen men and boys instantly, and another three were to die from their injuries. It was the worst colliery disaster in Burnley's long coalmining history. Two of those killed were aged just sixteen, a third was aged seventeen. The horror of that day has never been forgotten. Each year, on the anniversary of the event, a procession makes its way up Rossendale Road to a memorial situated just inside the cemetery gates, to pay their respects to those who perished. There were many acts of bravery that day from the miners themselves, the rescue workers from Boothtown Rescue Station, and one extraordinary woman, Sister Maud Waggott, the first aid nurse from Bank Hall Colliery who was one of the first people to go underground and gave aid. Many of those killed that day were brought out by means of the shaft rather than the new surface drift. Here rescue workers make their way down the pit, via the downcast shaft. (W. Rawstron Collection)

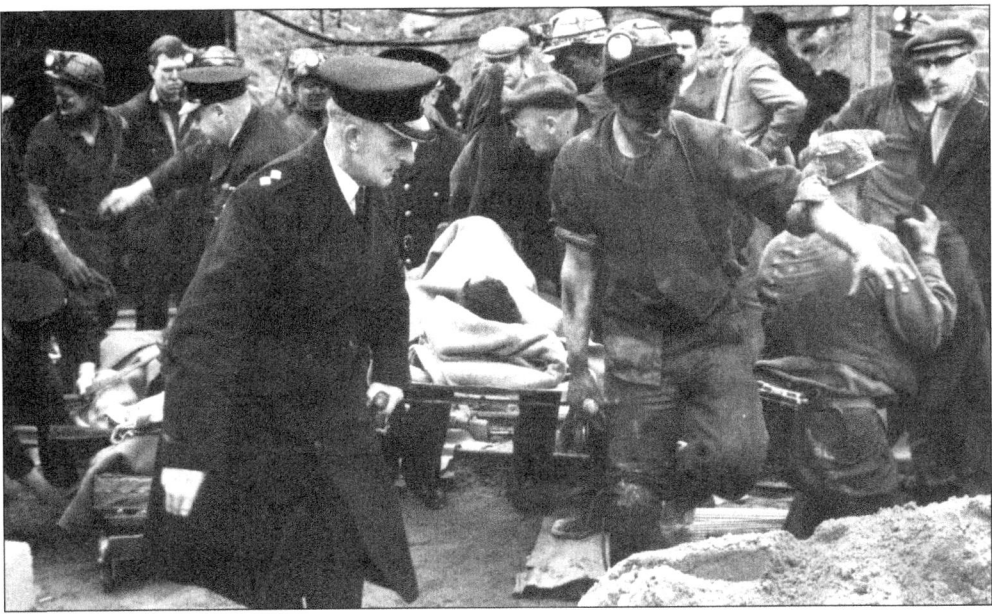

The frantic scene at the top of the surface drift at Hapton Valley Colliery on the day of the disaster, as rescue workers grapple to get through the crowds of anxious onlookers in order to get the injured out of the pit and into the waiting ambulances. Frightened relatives look on helplessly, hoping that their loved ones are among the survivors. Others look on in complete and utter disbelief. No words can express the feelings of those who were at, and in, the pit on that day, while heart-rending scenes such as these speak for themselves. It was, as the local newspaper put it, 'Burnley's Blackest Day'. Only acts of war and natural calamities have taken more lives than British coalmining over the centuries, something we might all choose to remember seeing this, the 'true price of coal'.

Mining in Union Seam at Hapton Valley Colliery was performed by the Universal Anderson Disk Shearer, seen here working on top of the armoured scraper conveyor. This worked by driving at the coal-seam with the revolving disk filled with carbon-tipped teeth. Following each cut, the conveyor was then moved forward into the space where the coal had been and the process repeated – the space behind was allowed to drop. Individual 'Dowty' hydraulic props were moved forward to support the roof, although later automatic hydraulic chocks were brought into use, as seen here. The Union Mine at Hapton Valley averaged about $3\frac{1}{2}$ ft in height and the length of the coalface was around 200yds. (Burnley Borough Council, Towneley Hall Art Gallery and Museum)

The mining of the Upper Mountain Mine at Hapton Valley, or the Top Bed, was begun in the mid-1960s and worked by the 'Plough', seen here. The Plough, full name the 'Roisshakonhobol Plough', was a German invention which worked exceptionally well in this seam by taking a thin slice of coal off in a number of passes. The Upper Mountain Mine only averaged around 32in in height. The 'Dowty' type of hydraulic prop can also be seen; these were 'pumped up' using a special key, or spanner. (Burnley Borough Council, Towneley Hall Art Gallery and Museum)

The monorail system for transporting materials at Hapton Valley was the latest method for moving tackle to the coalface, and is seen here with Barry Wilson obligingly 'posing' for the camera. Ideally they were fitted up by endless rope to a 'Pickrose', called a 'crab' down the pit, or an electric winch. In reality they rarely were, and had to be pushed all the way to the coalface using sheer muscle power rather than electricity. On the No.8 coalface in the Union Mine return gate where I was employed for a time, my mate Raymond Woods and I had to push these monorails uphill for many hundreds of yards. For this we dispensed with the coupling bar between, and used the blocks individually, having one block each. This wasn't too bad near the coalface, where the rails were comparatively level, but outbye they were more akin to the big dipper, as the weight on the arches buckled and bent both the arches and the rails. Going out though was a real joy; we'd put one leg through the chain, hang on to the shackle behind, and just let go, reaching speeds in excess of 30mph down the roadways. Later, my mate Goodie and I got the regular job at the bottom of the surface drift on the monorail taking materials through to the No.2 tailgate manrider – a real cushy number. We could get the manrider guard to bring us down a hot pie from the canteen for bait time, while the guard was always a soft touch for a chew of baccy – otherwise we'd take our time unloading the train. Blackmail, well, yeah! (Burnley Borough Council, Towneley Hall Art Gallery and Museum)

Above: Hill Top Colliery was another of the completely new drift mines planned by the NCB; this one was to be at Sharneyford, near Bacup, and was begun in August 1948. Here preliminary work is being started on the setting out of the new drift. This involved tidying up the exposed rock face ready for blasting. The pit is listed in the *Guide to the Coalfields* in 1951 as 'Still developing'. The coal to be mined was from the Union Mine and the manager was W. Dewhurst. (W. Rawstron Collection)

Right: Inside the drift at Hill Top Colliery work was ongoing driving the tunnels towards the seam of coal. Here in August 1950 the men are setting the steel arches as the tunnel advances using the basic materials at hand, including a wooden trestle to stand on and a primitive survey device. The two girders above the miners head on the trestle were named 'horse-heads' for a reason I never understood, and were slid forward and covered over in wooden boards to give at least some protection against falling debris at the working face of the tunnel. (W. Rawstron Collection)

New bunkers and a gantry were erected at Hoddlesden Colliery in 1957, with pithead baths being built at Hoddlesden Pit in 1953. Fireclay extraction for use at the pipe works at Hoddlesden ceased the previous year through lack of demand, and coalmining was then confined to the Upper and Lower Mountain Mines. Around this time special pit buses transported the men to and from the pit, saving the men the daily trek over the moors. On 29 September 1961 the colliery was closed down through lack of reserves and uneconomic workings. All that remains of this pit is the colliery access road below Pastures Higher Barn, now marked 'footpath'. During its last full year of production, the ninety-nine men at Hoddlesden Pit raised 17,743 tons of coal. (W. Rawstron Collection)

Massive investment was also carried out at Huncoat Colliery, near Accrington, by the NCB. The colliery is seen here from the pit's stockpile in the early 1950s. Prior to Nationalisation, the pit was owned by George Hargreaves Collieries, and was the first complete shaft sinking ever attempted by this firm, which later amalgamated with Hargreaves Collieries (Burnley Ltd). Many attribute the first shaft sinking by this firm as being the Scaitcliffe Colliery, but this firm only deepened the shafts there. (W. Rawstron Collection)

These old 3½cwt tubs, seen here at the upcast and coal-winding shaft, were soon to be replaced by larger-capacity 15cwt tubs. (W. Rawstron Collection)

The new larger-capacity 15cwt tubs operated under a completely new covered tub circuit. Full tubs raised up the pit ran around to be tipped into a tippler, which fed the coal directly up to the new screening plant at the pit. The whole operation was fully mechanised; empty tubs pushed forward with hydraulic rams discharged the full tubs out of the cages, which then ran under gravity controlled by 'wheel grippers', hydraulic devices that squeezed the wheels to the tippler. Empties went back around the circuit to the shaft and were hauled up a slight incline by creepers – and the whole process repeated. (W. Rawstron Collection)

The new screens and dirt bunkers can be seen here. In the background are the new rail bunkers to take the coal from Huncoat Colliery directly to the nearby Huncoat Power Station, the main customer for the pit. (W. Rawstron Collection)

The old wooden airlock can be seen here, along with the old headgear at the upcast shaft. In the background are the old coke oven chimneys. 'Jock' Smith poses for the cameraman besides the old $3\frac{1}{2}$cwt tubs and the new 15cwt tubs. Part of the old ginney that worked the surface layout can also be seen. The new airlock was completed along with the new headgear at the upcast shaft in April 1956. (W. Rawstron Collection)

At Huncoat Colliery 65hp diesel locomotives were installed underground for hauling the coal along to the shaft. Other improvements included a weighbridge installed near the shaft bottom. The pit worked the Upper Mountain Mine with longwall faces 150yds long using electric coal cutters and Sutcliffe gate conveyors. At one time all the coal from Scaitcliffe and Calder Colliery was raised at the Huncoat shaft. Following reorganisation, output from the colliery rose from 22cwt per man to 39cwt. The colliery was, however, closed on 9 February 1968 when the workable reserves became exhausted. Although the NUM's Joe Gormley said he'd fight against the closure if the men so wished, almost every man preferred to take the redundancy money. Little remains now of the former colliery, but an electric substation near the railway bridge that gave access to the pit is named 'Meadow Top Colliery Substation' which was apparently the local name for the pit. (W. Rawstron Collection)

Reedley Colliery at Burnley was adopted by the NCB as an experimental pit for the area. It is seen here with its back-to-back double-decked cages, which were superseded by the Sandvic Steel Belt Conveyor in 1956. This conveyor, the latest thing in mining at this time, took the coal through to Bank Hall Colliery via the underground drift that surfaced at Bank Hall, near the Eastern Avenue entrance, and weighed it at the same time. (W. Rawstron Collection)

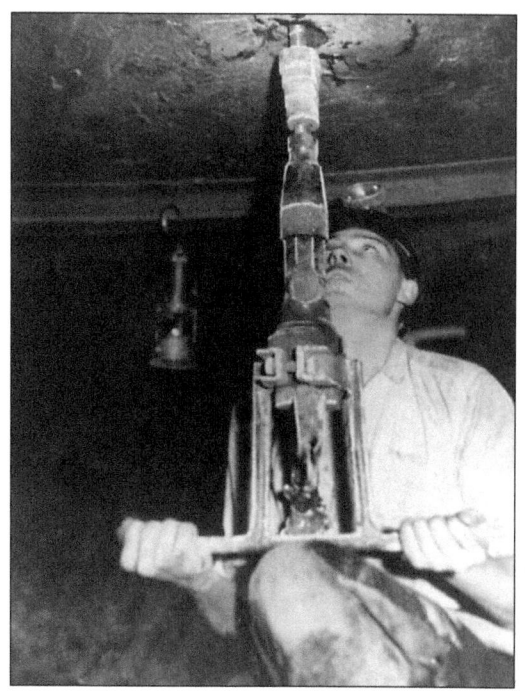

Roof Bolting being carried out at Reedley Colliery. Roof bolting was an American invention whereby the roof of the mine was drilled and a large anchor bolt, rather like a large Rawlbolt, was inserted. The theory was that the bolt would hold together the strata above – it was, however, disliked by the British coal miners, who preferred to have something they could see holding up the roof above them, be it timbers or girder-work. Fears over the safety of roof bolts were realised when six men were buried under a fall of a roof-bolted roadway at the Bilsthorpe Colliery in August 1993. Three of the men, including the colliery undermanager, were killed. (W. Rawstron Collection)

This view of a roof-bolted roadway in the Dandy Mine at Reedley Colliery shows why the miners were apprehensive over this system of roof support. The large span between the sides of the tunnel appears to have no supports whatsoever other than the heads of the roof bolts protruding through the roof. This system of roof support was, however, carried on at Reedley Colliery with great success. (W. Rawstron Collection)

A Joy-Loader, a kind of low-wheeled vehicle in use at Reedley Colliery, was adapted for use as a mobile-tracked vehicle for drilling roof bolts at the pit, apparently with some success. The 'mobile self-propelled and tracked roof bolting machine' as it was named, is seen here in 1950, and must have presented a frightening vision of modern mining, with from left to right: J. Smithson, overman; E. Holland, engineer; B. Duckworth, undermanager; A. Atkin, electrician; and T. Whalley, manager, all proud of their new invention. Reedley Colliery was closed down in May 1960. At this time the colliery employed 245 men, the total output in the last year of full production being 83,555 tons. (W. Rawstron Collection)

Salterford No.1 Drift was another of the new drift mines driven by the NCB during the acute coal shortage of 1948. The pit was driven to extract the coal beyond the economic reach of the old Towneley Colliery workings in the Dandy Mine, and, being driven in at 1 in 43, intercepted the coal-seam after 300yds. The pit in fact connected with the old Towneley workings, and water from Salterford No.1 was pumped into the old Boggart Bridge Colliery using a mono-pump working twelve hours a day. Here, looking up the main drift, we see the pit during driving. Notice the temporary conveyor belts which had been installed, and the unusual 'L' section track running besides the belt. The workings were limited towards the Brunshaw end due to the abandoned and flooded workings belonging to the old Rowley Colliery – consequently a 200yd barrier was maintained underground between the two pits. (W. Rawstron Collection)

By January 1951 the pit had, for the first time, produced over 2,000 tons of coal in one week. It regularly broke the output target and was allowed to 'fly the flag', as shown here. Compared with other local pits, which produced around 22cwt per man per shift, Salterford produced about 43cwt per man. The pit employed twenty-four colliers and 131 other underground men, including shotfirers, rippers, cutters, conveyor-belt-men, movers and maintenance men. Pay was good at the new pit – the local newspaper claimed that coalface workers could earn about £7.00 per week, while the surface men earnt about £5 10s. Salterford No.1, from the start a short-term colliery, was abandoned in April 1956, and work was started on the Salterford No.2 Pit in October 1953. (W. Rawstron Collection)

Salterford No.2 Pit, located on Foxstones Lane, is seen here on what appears to have been a snowy winter's day. The pit worked an area of the Dandy Mine not accessible from the No.1 Pit, even though it wasn't that far away from the latter. The Salterford No.2 Colliery employed 201 men underground and twenty-one surface workers. The manager was J. Malone, the undermanager J.G. Cartwright. There are no remains of the Salterford No.2 Colliery, which closed down in December 1959, as the whole site was landscaped. (W. Rawstron Collection)

The pithead baths at Scaitcliffe Colliery, Accrington, were opened on 22 May 1954 by R. Lowe, Area General Manager. (W. Rawstron Collection)

An 'interested' group of officials and workmen attended the opening – the person second left appears to be particularly bewildered by the whole occasion, as do a number of others! (W. Rawstron Collection)

The opening of the pithead baths at Scaitcliffe Colliery was followed by an inspection tour of the baths by the prospective customers, and other officials of the colliery. From their expressions, the workmen look as though they have just spotted a leak at the shower heads. (W. Rawstron Collection)

Opposite, above: The Thorny Bank Colliery Project, a plan to drive into the heart of Hambledon Hill to extract over 2.5 million tons of coal, was begun in 1950. The colliery was located at Hapton, on the Accrington side, were many of the surface buildings remain today. The pit was classed as a naked flame mine while it was being developed, and unfortunately in May 1951 there was an explosion in which a number of men were severely burnt. Peter Carr of Rawtenstall later died from the injuries he received in the blast. The new pit was intended to work the Upper and Lower Mountain Mines from a drift entrance besides the main Accrington to Burnley Road. Ventilation was achieved at the colliery by using the old West Side Drift belonging to the Hapton Valley Colliery, which was driven in the 1940s to prove the coal for that colliery beyond a fault. An installed fan here blew the air into the workings, rather than extracting the air, as was conventional. In this way the fumes from the diesel locomotives would not be drawn into the workings where men were employed. Here, around 1951, we see the new portal at the colliery and the battery locomotives used in driving the drift, along with the Jubilee side-tipping mine tubs used in driving. (W. Rawstron Collection)

Opposite, below: A group of unknown miners set the steel arches while driving the Thorny Bank Colliery in the early 1950s, using an Eimco bucket loader to help them. Coal production from four faces began in August 1952, mining from the Upper and Lower Mountain Mines. All the faces were equipped with power-loaders working double shifts. Coal was taken to the surface in 3-ton mine cars hauled by diesel locomotives. In 1959, the colliery was employing 383 men underground and thirty-five surface workers, the manager at this time was J. Yuill, undermanager G. Turfrey, both of whom later when to Hapton Valley Colliery. (W. Rawstron Collection)

A general view of Thorny Bank Colliery taken from the top of the screens and gantry looking towards Huncoat and Accrington. The colliery was constantly in the news for breaking production records. In 1953 over 80,000 tons of coal was raised at the pit, and the following year it was expected that 100,000 tons would be raised – and it was. (W. Rawstron Collection)

Opposite, above: Using the most modern methods of mining available at the time, Thorny Bank produced vast quantities of urgently-needed coal following the Second World War. Here the coal cutter is in use in the mid-1950s, tearing out the coal from the Lower Mountain Mines. It was estimated that this mine, along with the Upper Mountain Mine, would produce about 4,500,000 tons of coal, and that the pit would have a life of some thirty-odd years. However, work finished at the colliery on 12 July 1968 due to adverse geological conditions and, some might add, political motives. Two years before, the colliery was producing 160,000 tons of saleable coal per year at an overall productivity level of 44cwt-per-man shift. But in its last full year of production the pit produced just 138,344 tons, with a manpower of 307 men. (W. Rawstron Collection)

Opposite, below: Towneley Colliery around the 1930s, showing the pit's ginney, which passed over Rock Lane to make its way to the canal side near Finsley Gate Wharf. The continuation of the ginney track behind the cameraman was the scene of an infamous murder back on the last day of 1897. The victim, John Keirby Pickup, a colliery banksman, had sacked fellow pit employee James William Howe and Howe was forced to find employment at the Reedley Colliery, a longer distance away and on a much-decreased wage. Five weeks later, Howe sought revenge. In court he said he'd attacked Pickup with a stick, but evidence showed he'd used a lantern and that his clogs were covered in blood and hair. Pickup died from his injuries on 10 January and Howe was hanged on 22 February 1898. (Peter Nadin)

TOWNLEY COLLIERY, BURNLEY. Copyright.

The Towneley Colliery was one of the first local pits to be closed down under the then new NCB; its long and diverse underground haulage being the main reason for its closure. A manrider shaft sunk on Broad Ing in or around 1907 saved a great deal of travelling time for the men, but the coal still had to be hauled long distances underground. There was still plenty of coal to be extracted, but this would be done by driving new drift mines near Red Lees, to be named Salterford No.1 and No.2. The pit is shown here in 1949, just prior to the demolishment of the surface buildings. At the time of closure, the colliery, which was mining from the Dandy, Yard and Lower Mountain Mines, employed 295 men underground and ninety-five surface workers. While the pit was being salvaged, a huge girder from a road junction was brought to the surface. The girder, weighing in at one ton, was originally part of the famous great Ferris wheel at Blackpool. It was placed on top of the cage with a man standing besides it to steady it – as it was too large to fit in the cage. The colliery site was eventually cleared, and the shafts filled – but not properly. In 1986 one of the old shafts opened up again as the fill slumped. Today, the site of the old pit is covered with modern housing. (W. Rawstron Collection)

Four
The Ginney Systems Around Burnley

At Burnley coal mines many systems of surface haulage, along with underground haulage, were operated on what became known as the 'Burnley System' or, locally, the ginney system. The system was allegedly invented by one of the Landless family at the Brierfield Colliery, known locally as Marsden Pit. On the surface, these were in effect miniature double-tracked railways, worked by an endless chain with a wheel at either end. Many of these systems worked in and around Burnley connecting the pits with canal-side wharfs or the railways for the coal to be transported further afield. One main disadvantage was that these ginney tracks had to work on a level plane or evenly graded track. Consequently many miles of embankments, cuttings and trellis bridges had to be constructed to take the ginney on its level course. An advantage was that the ginney could be operated with just a small engine, the tubs themselves actually taking the weight of the long lengths of chains used to operate the ginney. Any change in direction was achieved by the use of 'turning blocks', which were large stone or brick buttresses that carried a turn-wheel arrangement to deflect the course of the endless chain. A few of these still survive in the Burnley Coalfield. One of these ginney belonged to Rowley Colliery, and here we can see the gearing and other arrangements at the ginney head, that portion of the ginney nearest the pithead, as it was at Rowley around 1867. Chain-winding in the shaft to raise the coal was also used at a number of the local collieries, namely at Rowley, Hapton Valley, Cornfield and Gannow Pits.

From the pithead at Rowley Colliery, the ginney went straight across what was then open fields, crossing, as it did, embankments and bridge works to the canal-side near the present-day 'Culvert'. Here the coals could be unloaded into barges for transportation throughout Lancashire. It was stated that over 400 tons per day could be sent to the canal wharf using the Rowley Colliery ginney system, at a speed of just over 3mph. The tubs were usually set 25yds apart. The wheels at both ends of the ginney were usually 3ft in diameter, and the cost per day to operate the ginney at Rowley Pit amounted to just 6s and 7d. The total length of the colliery's ginney, in its direct line with the canal at Yorkshire Street, was 1,980yds. The ginney from the Bee-Hole Colliery at Brunshaw Bottom connected with the ginney from Rowley Pit at a midway point named 'The Turning'. From this point coal from Bee-Hole, or for that matter Rowley Pit, could be sent to Bank Hall Colliery for screening. (Picture, Proceedings No.E.1 of *Mining Engineer*, 1867-68)

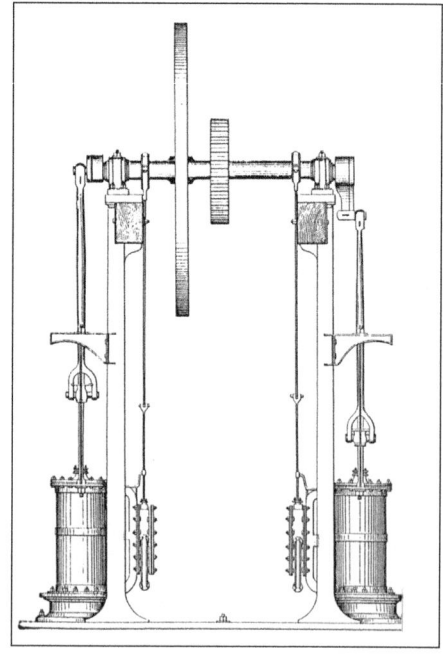

The ginney system was used at Burnley, and to an extent in the Rossendale Valley Pits, primarily because it was economical. Once the initial layout of the track was organised – taking out the cuttings, putting in embankments and building wooden trellis bridges – the cost of running thereafter was minimal. This was due to the small steam engines required to work the system, the engine at Rowley Colliery using just over 5hp. The Rowley Colliery ginney engine, seen here, was similar to the engine used at Hapton Valley Colliery, although their engine used 18.80hp. This was necessary, though, as it was used to operate a 3,207yd-long ginney – probably the longest ginney in the Burnley Coalfield. Another advantage of the ginney system was that the tubs 'automatically' attached and detached themselves to the chain, which simply dropped into a slot on top of each tub. (Picture, Proceedings No.E.1 of *Mining Engineer*, 1867-68)

The ginney system that ran from Bee-Hole and Rowley Pits in operation in May 1927. The tub has just emerged from a tunnel that ran under Queen Victoria Road, constructed in that year. From here, the coal tubs were taken over the River Brun on yet another wooden or steel trellis bridge, to Bank Hall Pit. This ginney was probably abandoned when the last tub of coal was raised at Rowley Colliery on 8 May 1928, although it may have continued to operate from the Bee-Hole Pit until that closed. Ginney systems operated at other local pits, including Gannow, Clifton, Burnt Hills, Gambleside, Habergham, Railway and Copy Pits. (W. Rawstron Collection)

Hapton Valley Colliery, having no direct links with canal or railway, had an extensive ginney system. One ginney ran over the moors from behind the Porters-Gate Colliery for almost two miles to connect with the ginney at Hapton Valley, before running on to either the Smallshaw Railway Sidings or the canal wharf at Gannow. The latter section of ginney track ran until at least the early 1950s, when the Hapton Valley pit was reorganised by the NCB – coal was then taken from the pit by Coal Board wagons. This view of the ginney head at Hapton Valley during reorganisation shows the method of the tubs' attachment to the ginney chain. Notice the 'L' section rails, and the disk-type wheels used extensively throughout the Burnley Coalfield on the ginney tracks. Notice too, the steel lattice headgear soon to be replaced by a girder section headgear. (W. Rawstron Collection)

The old Padiham Coal Staith, where the ginney system finally arrives from the Cornfield and Grove Lane Collieries for distribution of coal to the mills and household customers at Padiham. Much of the coal, however, only travelled a few yards to the gasworks, which was situated on Station Road up to the early twentieth century. The two men either side are obviously workmen employed at the coal staith, while the man in the middle appears to be more important, possibly the foreman at the staith. Here we see the size of the tubs used on the ginney: notice the handholds on the tubs for manipulating the tubs both at the ginney and at the staith itself. Following the closure of the Cornfield Colliery, the ginney track fell into disuse, as did the Padiham Coal Staith, in or around 1936. (Duncan Armstrong)

Five

People and Places

Following Nationalisation of the British coal industry, the NCB saw the need for training for new recruits – both from a safety aspect and a production aspect. Training was carried out at Bank Hall Colliery, the area's largest pit, and at the Municipal College. Six weeks training was the norm: alternately one week at college and one week at the pit. Here a group of the area training staff are at what appears to be the main entrance to Burnley Municipal College, probably around the early 1950s. Back row, from left: J. Tuck, training officer; E. Forrester, coalface instructor; M. Brown, instructor; W. Crowther, coalface instructor; E. Radcliffe, coalface instructor. Middle row, from left : H. Brennand, instructor; W. Tong, training instructor; H. Pounder, coalface instructor; H. Hindle, coalface instructor; R. Wilkinson, coalface instructor; N. Walton, training officer; T. Shepherd, coalface instructor; J. Wilson coalface instructor. Front row, from left: G.W. Lord, instructor; P. Littler, chief instructor; A. Mitchell, training officer; A.B. Wilkinson; head of mining department, Municipal College; T.I. Jeremiah, area training and education officer; W.E. Rawstron, manager of Bank Hall Colliery; E. Lord, training officer; J. Large, coalface instructor; D. Frankland, coalface instructor. (W. Rawstron Collection)

The new area training rooms for the practical and theoretical side of coalmining were opened at Bank Hall Colliery in June 1953. Here subjects such as first aid, safety underground, mine gasses and ventilation were taught. The newly recruited miners might have the chance to descend Bank Hall Pit by the No.1 shaft, where training was also given on a 'mock coalface' under the supervision of a training instructor. Seen at the opening of the new training rooms are, left to right: A.B. Wilkinson and J. Holmes, welfare officers; H.E. Clegg, area production manager for the NCB's North Western Division; T.I. Jeremiah, head of training at Bank Hall Colliery; H.C. Riley, planning; W.E. Rawstron, manager at Bank Hall Colliery. (W. Rawstron Collection)

Opposite, above: Altham Coke Works was built on the site of the old Moorfield Colliery, which was often called 'Dicky Brig Pit' after the nearby Pilkington Bridge. The bridge itself was named after a farmer called Richard Pilkington, whose farm was nearby. Here, a group of the Altham Coke Works' fire-fighting team line up in an orderly fashion, *c.*1960. Left to right are: George Livesey; John Blades; Jack Feltell; Joe Barnes; Brian Addison; Ron Jackson; John Czyneck (Hungarian refugee). The unidentified gentlemen with the suit was the local area supervisor. (Brian Addison)

Opposite, below: On 7 March 1953, a number of long-service awards were handed out to miners with fifty years service in the industry at the old Hall Inn at Burnley. Many will recall the Hall Inn (demolished in the early 1960s) at the top of Hall Street near the present-day Keirby Hotel. Here we see the presentations being made. T. Whalley, manager at the Reedley Colliery, hands over a certificate to H. Dewhurst. Others present are, left to right: G.W. Bottomley; H. Smith; A. Duckworth; T. Quinn; F. Kellet; W. Hawke. (W. Rawstron Collection)

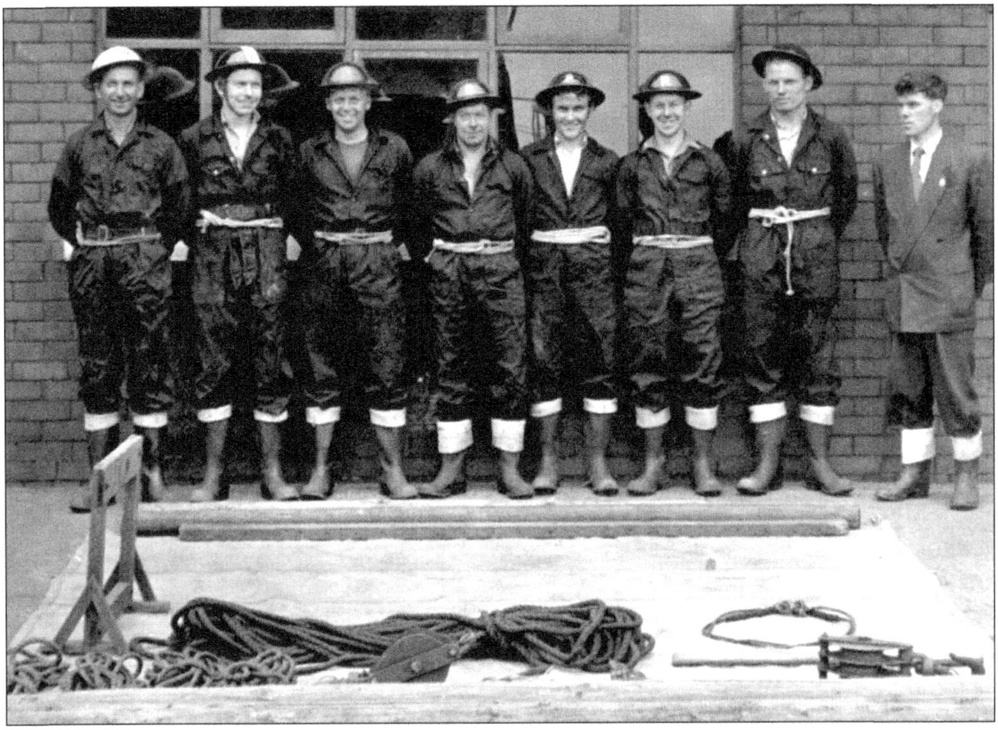

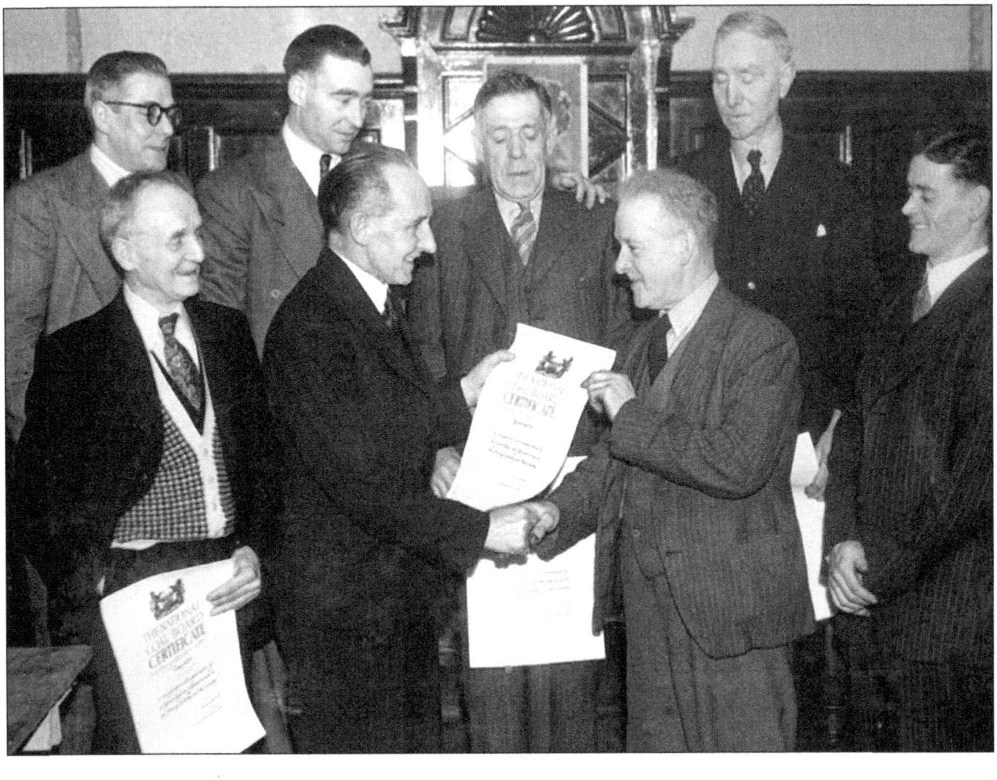
THE NATIONAL COAL BOARD
CERTIFICATE

An example of the type of long-service certificate the men would have received at the Hall Inn, only here the award is made out to Thomas Poole, a Bank Hall miner. The document is signed by Lord Robens of Woldingham, chairman of the NCB in the 1960s. Lord Robens reshaped the industry into an industrial power source for the '60s and '70s, albeit through scores of colliery closures. As a Union official, a Labour MP, and a post-war Labour minister, he was a high-powered industrialist. He was, however, bitterly despised in South Wales for his role over the Aberfan disaster. On 21 October 1966, a mountain of black sludge and colliery waste slid down from a Coal Board tip, burying the Pantglas School at Aberfan and killing 116 children and twenty-eight adults. Lord Robens was informed of the disaster, but decided to keep his other appointment that day, that of being installed as Chancellor of Surrey University, only going to Aberfan the following day. As local miners and rescuers were working throughout the night pulling out dead children from the muck and slime, their boss was pulling on his ceremonial robes in Guildford. The tribunal enquiry after the disaster found the Coal Board guilty of neglect, but the admission of responsibility was dragged out by the NCB. Lord Robens did his utmost to ensure that the NCB did not have to pay for the removal of the tip following the disaster. The disaster fund itself, which was intended to pay for the funeral expenses of the dead infants, had to pay £150,000 towards removing the NCB's tip from the NCB's land. This wrong was finally righted many years later in 1997 by Ron Davies, one of the few Secretaries of State who was both Welsh and understood the deep anger that the NCB's behaviour had provoked in the Welsh Valleys. Alf Robens lived until he was eighty-eight, a great deal longer than the 116 children who perished in Pantglas School that sad day. Greater disasters have occurred in British Coalfields as far as loss of life was concerned, but surely the Aberfan Disaster, as well as having occurred within living memory, was the most heart-rending in its consequences. Compassion, it would appear, did not extend to the government-run NCB, who had said on Nationalisation that 'The pits are now run by the Government on behalf of the people'. In the same circumstances, I'm sure the 'people' would have acted otherwise and with more understanding. (Keith Poole, son of Tommy Poole)

The long-service award was presented to Tommy Poole when he finished at Bank Hall Colliery on the last day of April 1969. At 2.25 p.m. these three men stepped out of the cage at the colliery for the last time – thus ending their mining career. They were taking part in the new redundancy scheme at Bank Hall, recently introduced as part of the recruitment drive, and seventeen other miners at the pit also took early redundancy. Left is Tommy Poole, fifty-six, a miner for forty years, having started at Towneley Pit and moved to Bank Hall in 1934. He was secretary of the Union here for almost ten years. Next is Joe Pepper, fifty-five, the longest-serving miner at the pit, having worked at Bank Hall for forty years. The third man is Billy Warne, who'd been at Bank Hall just over three years. Prior to that he worked at Towneley Park Pit, Salterford No.1 and No.2 Pits, Copy Pit and Hill Top Colliery. The men were congratulated on their long service by the pit's undermanager, James Wormwell. (Keith Poole, son of Tommy Poole)

Burnley's best known local historian, Ken Spencer, aged eighteen in 1947, outside the back door of 18 St Matthew's Street, Burnley, in his 'pit clothes'. These, in those days, consisted of old army gear obtained from the local Army and Navy Surplus Stores. Ken was a coalmining optant, not the same as a Bevin Boy. The Bevin Boys were selected by ballot to work in the coal mines; one in nine of all those who were called up for National Service had no option but to go and work in the pits. An optant, such as Ken, chose to work in the local coal mines. Notice the bait box under Ken's arm, usually packed with jam butties. These tin boxes had to be used underground to stop the pit mice and rats (which went in to the pit with timber supplies from the stockyard) getting at your food while you were working. The rodents were very adept however, and it wasn't unknown for them to figure out how to remove any loose lids on the boxes. Miners who took their bait just wrapped up in bread paper, would stuff this down their coat sleeves and then hang the coat high up in the roof of the tunnel. The colliers often got a nasty nip when they went for their food at bait time – for pit mice became very agile at scaling pit props. (Ken Spencer)

The Jobling family were associated with the Cliviger Coal Co. for generations. John Jobling, pictured here at the Union Colliery in 1921, was the manager for over forty years. It was John Jobling, as chairman, who persuaded the Burnley Rural District Council to construct a reservoir above Mereclough to supply water to the village in 1904. However, it might also be pointed out that the springs, which normally fed the village, were fast disappearing because of the Cliviger Coal Co.'s mining activities. In the same week in which John Jobling died, two local ladies had a remarkable escape when they fell into a hole, which suddenly appeared through subsidence, while they were crossing the recreation ground above the Railway Colliery. John Jobling was on numerous committees and public bodies: he was a magistrate, president of the Cliviger Brass Band, and a director of the Old Brewery at Cliviger. (W. Rawstron Collection)

The second annual miners dinner was held at the Hall Inn, Burnley, on 27 March 1954 – I'm afraid I don't know where the first annual miners' dinner was held! Here we see a number of miners and management from the local collieries at this, the second 'do'. Back row, from left: J. Dean, treasurer; J. Hudson; W. Spencer, secretary; H. Forest, miners' delegate. Middle row, from left: R. Whitehead; T. Brown, undermanager at Reedley Colliery; B. Kennedy, undermanager at Fence Colliery; A. Moore. Front row, from left: B. Gilbert; W. Hawke, chairman; J. Ashton, B.E.M.; T. Whalley, manager at Reedley Colliery; W. Rock; H. Smith. (W. Rawstron Collection)

Albert Woodward, left, and Henry Ormerod at the Salterford No.1 pithead baths, mid-1950s. Albert Woodward had a remarkable career in local coalmining. He started at the old Habergham Colliery in 1941 and worked there until it closed in August that same year. From there he went to Wood End Pit, Towneley Park Pit, Salterford Pit, Copy Pit, Deerplay and finally to Hapton Valley Pit. 'He's worked at seven pits, and closed them all,' his wife Jean quipped in, with a smile. Albert completed forty-two years in the local coal mines and took retirement when the area's last pit, Hapton Valley, closed down in 1982. (Albert Woodward)

The National Coal Industry Social Welfare Organisation (NCISWO) had, since its inception, spent £440,000 in Lancashire on recreation facilities such as clubs, institutes, sports grounds, playing fields and pavilions. Just two and a half years before the Bank Hall Social Club was built, the land was a former colliery waste site for Bank Hall Pit. The building itself consisted of the main hall, concert room (seen here), billiards, recreation and committee rooms. The club survived the closure of the colliery with which it was associated, and was for many years a popular entertainment centre in town. However, in the early hours of Friday 2 July 1993 the club was gutted by fire in a suspected arson attack. Only the shell of the former club remained and this was later demolished. Covenants in the deeds to the land prohibit it from being used for any other purpose, and today the site of the former club is derelict wasteland. (W. Rawstron Collection)

Opposite, below: The new playing fields that were adjacent to the Bank Hall Social Club were opened on 2 August 1955, two years before the club itself opened. They were just one of the 'recreation facilities' opened by the NCISWO. The opening ceremony is seen here, the only person identified being W.E. Rawstron, fourth from right, chairman of the Bank Hall Welfare Committee and manager at Bank Hall Colliery. Covenants in the deeds prohibit the land from being used for any other purpose, and today the playing fields are home to Worsthorne Football Club. Burnley Caving Club utilise one of the former shower blocks at the playing fields as a club room, and another room has a climbing wall and ladder-work practice utility for prospective new members. (W. Rawstron Collection)

Retired miners at Bank Hall Colliery often attended the annual dinner held at Bank Hall Social Club. On this occasion, in May 1969, there were over 180 retired miners in attendance. The guests enjoyed a buffet supper and listened to concerts by local artists as well as being given a token worth the princely sum of £1 to buy refreshments in the club. Here, president of the Management Committee, B. Conner, presents one of the retired miners, J. Riley, with his £1 token. Looking on, left to right, are B. Bowes, K. Bowes, J. McAlleese, J. Kirkham and E. Goulding. (Photo courtesy of Mrs Thomas, daughter of Mr B. Bowes)

Almost twelve months after the opening of the Bank Hall playing fields, on 16 June 1956 the new recreation and sports centre that would cater for the 1,100 men employed at Bank Hall Colliery was opened by Edwin Hall, North West General Secretary of the NUM. Cllr L.K. Crossley BEM, Burnley Area Miners Agent, said at the opening ceremony that: 'Bank Hall had toyed and played for over thirty years for such facilities, and now that we have achieved this, I would like to pay tribute to the men at Bank Hall who had a shilling a week stopped out of their wages to fund the place.' The tour of inspection of the new pavilion and greens includes Mr Hall, fourth from left, while on his right, front row, is Cllr L.K. Crossley. On Mr Hall's left is R. Lowe (wearing a trilby), Burnley Area General Manager. The proceedings concluded with Mr Hall sending a 'jack' down the new bowling green to mark the first official game on the new facilities. (W. Rawstron Collection)

Opposite, below: 'I've never been down a mine before, I think everything is marvellous,' said eighteen-year-old Miss Turid of Oslo, Norway, who was about to make the descent down Bank Hall Colliery shaft on 19 July 1949. Miss Turid was the guest of H.E. Randall, MP for the Clitheroe Division, and his wife during her six-week stay in Britain. Dressed like any other miner, complete with heavy boots, and her fair hair tucked neatly under her steel safety helmet, she stepped into the cage for the 1,500ft drop to the pit bottom. Besides descending Bank Hall, during her stay in Britain Miss Turid visited an approved school, the Stratford-on-Avon Memorial Theatre, Canterbury, Edinburgh and Glasgow. F. Sharples, ventilation officer at Bank Hall, shows Miss Turid how to use her pit lamp prior to the descent. On the left is H. Randall. (W. Rawstron Collection)

'I was dumb-founded', said John William Banks of Burnley, when he received a letter in January 1953 from the Prime Minister informing him he had been awarded the BEM for services rendered to the coal industry. Chairman of Bank Hall Sports Club, Mr Banks commenced coalmining work in 1896 aged twelve as a drawer at the Hoddlesdon Colliery, near Darwen, becoming a coal-getter four years later, aged sixteen. When he first came to Burnley in 1905 he began work at the Rowley Colliery before going to Bee-Hole Colliery, where he worked for some time as a fireman, before resuming work as a collier because it paid more money. His days at the coalface came to an end when when a pick was accidentally driven through his elbow. He had to give up pit work for nearly two years, undergoing no less than five operations on his injury. The family income for that time was just 30s (£1.50) a week, and with this Mr and Mrs Banks had to bring up eight children as best they could. He began work again at Bank Hall Pit in 1927 as a night watchman; the injury prevented him from working underground or doing heavy duties. In order to make ends meet he was obliged to work seven days a week for ten years, forsaking all holidays, even working on Christmas and New Year's day. Later he  became a banksman at the colliery, in charge of the cages at the top of the shaft, and shortly before his award he announced he was going to retire. The management asked him to continue working, and he was found work as a switchboard operator. Mr Banks also served with the Royal Field Artillery during the First World War and with the Burnley Home Guard during the Second, and here he proudly shows off his medals. (W. Rawstron Collection)

Previously the Director General of the BBC, Mr Foot resigned that position to become chairman of the Mining Association, embarking on a three-month tour of the British Coalfields. He began his tour in Lancashire at Bank Hall Colliery on 31 May 1944, seen here, afterwards visiting the Huncoat Pit. Mr Foot said that: 'my aim was to get to grips with problems and endeavour to bring about a state of security for labour, capital and the industry.' The picture is important, for it shows members of the Burnley area pit management teams prior to Nationalisation. Left to right: Col. T.M. Brooks, director of Towneley Coal & Fireclay Co.; Col. G.G.H. Bolton, Lancashire Association Collieries; Tom Lund, managing director of Hargreaves Collieries Ltd; Sir Robert Burrows, chairman of Manchester Collieries; G.C.M. Barlow, chairman of directors, Hargreaves Collieries Ltd; Robert Foot, representative, Coal Owners Association; W.E. Rawstron, manager of Bank Hall Colliery; H.E. Clegg, agent, Hargreaves Collieries Ltd. (W. Rawstron Collection)

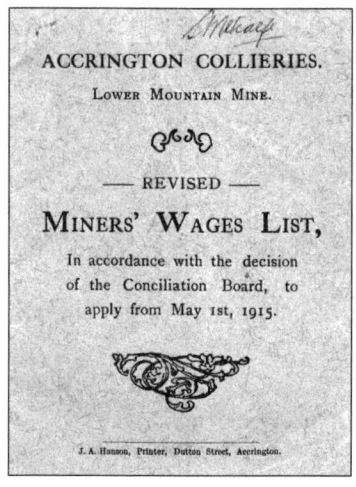

Accrington Collieries' Lower Mountain Mine Revised Wage List, in accordance with the decision of the Conciliation Board, to apply from 1 May 1915. At this time banksmen aged thirteen to fourteen were paid the basic wage of 1s 6d. This very important and responsible job involved being in charge of the pit cages at the top of the shaft, loading and unloading the cages. Drawers (those taking the coal to the shaft in tubs) and ginney tenters were paid the base rate of 2s 6d. Colliers working pillars not more than 8yds across were paid 2s 4½d per ton. 'War Bonus of 16% and War Wage of 1s. 6d. per day for persons over sixteen years of age, and 9d., per day for persons under 16 years of age is added to the above prices', the booklet concluded.

Over thirty years after the closure of Burnley's largest pit, Bank Hall, the name Albert Dugan is still synonymous with the NUM at the pit with dedication to helping others. Those injured at the pit, their wives, and the widows and children of those killed, will all remember Albert for his work in getting compensation for the injuries and deaths at the colliery. His commitment to those unfortunates often went far beyond the call of duty of ordinary Union men. Albert Dugan, born on 27 February 1922, entered the mining industry in 1936. He returned to coalmining after serving in the RAF during the Second World War, getting involved with the NUM in 1946. His ability to put forward an argument for compensation soon made its mark on senior management at the colliery. Through the quality of his debates, Albert was soon recognised by the Lancashire branch office of the NUM at Bolton. He became the Bank Hall Colliery Union delegate at conference, was elected as the Lancashire, Cumbria and North Wales executive, and also became the delegate of the NUM at the Trade Union Conference. Once, while on a visit to the Miners' Convalescent Home at Blackpool, several members of the Union were being shown by a senior member of staff the exact spot on the carpet where the Prince of Wales had stood while opening the home. Albert burst in through the group and blasted out, 'Never mind where the bloody Prince O' Wales stood, what about these Burnley lads who were locked out of the home last night?' At this time residents had to be in before 10.00 p.m. or they were locked out. Albert then added, 'I'm going to get these silly bloody times altered,' and he did. I can't verify the following, but I'm assured it did happen. One day a miner was leaving Bank Hall when, by all accounts, he had a heart attack and died in the street. Albert Dugan was soon on the scene and said, 'Quick, drag him [the deceased] back on to the pit top before anyone else sees him.' 'How's that?' asked one of the bystanders. 'Because,' said Albert, 'if he deed [sic] on pit top, then at least his wife and kids will get some compensation.' When the Bank Hall Colliery closed down in 1971, Albert found employment at the Lucas Works, where he continued to fight for workers' rights, dismissals and compensation claims. Albert sadly died on 14 January 2002. His funeral was a fitting and final tribute to a great man, one who dedicated his whole life to helping others in need. (Irene Dugan)

The Miners' Convalescent Home, opened by the Prince of Wales in June 1927, was built at a cost of £160,000, including the equipment and beds for 132 men. Arthur Scargill planned to save the seafront home at Blackpool when it was put up for sale in June 1996, but was shot down. Trustees and management committee members of the Home turned down a bid by the Yorkshire area NUM to buy the home, with payment spread over a five-year period, and chose instead to sell the home to outsiders for cash up-front. The majority of NUM-nominated trustees and management committee voted along with the British Coal-nominated trustees and committee for a cash bid. Every miner in Lancashire paid a weekly contribution to the upkeep and running of the Miners' Convalescent Home and the decision to sell caused a great deal of controversy.

Billy Cook started work at Bank Hall Colliery in 1947 just after Nationalisation, at the age of twenty-eight. He remained there until it was closed down because of the frequent ignitions of gas at the coalface, and was transferred to Hapton Valley Colliery. His strong argumentative talent soon found him a place on the local branch of the NUM, and later still he was appointed to president of the Burnley branch of the NUM. He was a member of the local Labour Party executive committee, and even once stood as a candidate in a local council election. Billy retired when Hapton Valley closed down in 1982 and because of ill health. Billy died on his birthday, 8 October 1987, and the town's MP, Peter Pike, paid this respect to him: 'Billy Cook was a tremendously loyal member of the Labour Party and a dedicated worker in the trade union movement.' (Sylvia Marsden, Billy Cook's stepdaughter)

Photographs of the old Moorfield Colliery are rare, so I was pleased when handed this late 1910s or early 1920s image of a group of miners at the colliery following a plea in the local newspapers. Besides the miners at the pit, complete with their flame safety lamps, there are a number of women shown who probably worked at the pit top screening coal. The man with the boater hat on the right-hand-side was possibly connected with the colliery management. The photograph also shows George Howarth Metcalf, far left and inset. Born in 1886, Metcalf started work at the Moorfield Colliery as a drawer, working his way up to collier. Following an accident at the pit in which he lost an eye – for which he received £125 – he was laid-off work for some time. When he recovered he became the Check Weighman at the Martholme Colliery at Great Harwood, later getting the same position back at Moorfield. By all accounts George's sister is one of the women in the picture. (Kevin Metcalf)

On 19 August 1949 it was the turn of Miss New Zealand to descend a coal mine, this time at the Huncoat Colliery. Miss Woodward was on a six-week tour of Britain and was accompanied by Second Officer Peat of the WRAFS, herself a New Zealand woman. They were met at the colliery by John Whittaker, the pit manager; J.M. Holmes, Labour Welfare Officer; and W. Oldroyd, the colliery undermanager. The underground tour took about an hour and a half and was followed by a lamb meal in the pit canteen. Other engagements taken by Miss New Zealand included a visit to Ambleside in the Lake District and a tour around the Blackpool illuminations. Notice the old three-deck cage then in use at the Huncoat Colliery, and the small tubs. (W. Rawstron Collection)

Ormerod House was the home of the Thursby family, notably the Revd William Thursby who married Eleanor Hargreaves, of Hargreaves Collieries. They arrived at Ormerod House in 1834 along with their young family, four sons and one daughter. A second daughter and an additional three young sons were to be born at Ormerod. Due to failing health, Mr and Mrs Thursby moved to Brighton in 1869 and their son and heir, John Hardy Thursby, took over both the Burnley coal concern and the Ormerod Estate. This estate was put up for sale at the Bull Hotel in Burnley in September 1922. Ironically, Ormerod House itself had to be demolished in 1947, after the Cliviger Coal Co. undermined the house while working in the Arley Mine. (Burnley Borough Council, Towneley Hall Art Gallery and Museum)

Bank Hall, Burnley, was purchased when it was still a half-timbered building by Revd John Hargreaves soon after his marriage to the widow of Henry Blackmore of Fulledge, thus inheriting the estates and coalmining rights there. When the Revd Hargreaves died, the estates and coal mines went to his nephews, John and James Hargreaves, and so came into being the 'Exors of John Hargreaves', the great Burnley coal masters. Later, and again through marriage, this time to John Hargreaves' daughter, Charlotte Ann, Bank Hall became the home of Gen. James Yorke Scarlett, our local hero. During the First World War, Bank Hall became a military hospital, and in 1916 it was offered for sale to the Burnley Council who later converted it to a maternity hospital. In October 1959 parts of the old Bank Hall were deemed unsafe due to subsidence and the building was demolished. Today, a residential nursing home takes the site of the former Bank Hall. (Peter Nadin)

There were of course many disputes and strikes in the coal industry. This image is thought to be of the 1972 miners' strike called on 9 January. By 16 February there were total electricity blackouts lasting up to nine hours at a time and industry was reduced to a three-day working week. This, and other strikes in this year, brought down Ted Heath's Tory Government – but they sought, and got, their revenge under Margaret Thatcher, and later John Major, when the government decimated the coal industry in 1992. Mrs Thatcher claimed to have 'beat the miners' in their strike of 1985, but at a terrible cost. Billions of pounds in lost production, millions in police man-hours while policing striking miners, and thousands of miners' jobs lost forever. The strike of '85 was not about wages, but about saving pits – Scargill was right when he said that in less than ten years time there would only be half a dozen pits in Britain. Here we see Billy Cook, Union man at Hapton Valley Colliery, in a defiant mood pointing to the senior police officer. To his left is John Barker, and behind him Harold Durant. Top right is Herbert Whitehead, with glasses and mopping his brow. By all accounts Billy Cook was later arrested for throwing an egg at Prime Minister Ted Heath. (Sylvia Marsden, Billy Cook's stepdaughter)

Opposite, above: When Fred Cater, seen here as a twenty-nine-year-old miner in the early 1950s, told his workmates on the nightshift at Salterford Colliery that he had decided to leave Burnley to start a new career in Australia, they decided to give him something to remember them by. A number of the nightshift held a party in Burnley's General Havelock Hotel, and John Wilson handed to Mr Carter a cigarette case and lighter on behalf of the men. More coal was cut in the taproom of the local pub over a game of Don than was ever mined at the pit, it's been said. This is especially true in the pubs around the Wood Top Districts of Burnley and those pubs in the Duke Bar area. 'Have you moved pans over?' or 'Have you cleared that ripping?' were common greetings to men coming in to the pubs. (Albert Woodward)

Opposite, below: A group of officials, firemen and overmen at the Thorny Bank Colliery at Hapton, near Burnley, c.1962. The men are in typical pit clothes; any old jacket and pants would be used at this time. The NCB issued bright orange boiler suits around the mid to late 1970s, but by then Thorny Bank had closed. The men are, from left: Noel Neeley, who appears to have tucked his bait down his pullover, as was common; Fred Sowerby; George Thursden; Jack Smithson; Terry Long; Jack Hudson. Terry Long is carrying in his pocket a flask which, strictly speaking, wasn't allowed as the inside of the flask contained glass. However, it didn't stop the practice, although most of the men took water into the pit in aluminium or plastic bottles. (W. Rawstron Collection)

109

Henry Johnson started work at the Bank Hall Colliery at Burnley in January 1939, following his father who also worked there. Henry later went to the Scaitcliffe Colliery at Accrington, where in 1952 he was nominated as 'Miner of the Year' – quite why I'm afraid neither I or Mrs Peel know. It is known that the NCB gave out awards for all sorts of 'achievements', ranging from good attendance, so many days without an accident, and even having polished your pit boots! Henry, who later went to the Thorny Bank Colliery and later still to William Blyth's Chemical Works near Accrington, sadly died in March 1980. (Mrs D. Peel, Accrington)

Right: All I know about this photograph is that it was captioned 'Presentation to fire-fighting team at Thorny Bank Colliery.' Perhaps readers might be able to put the names to the four people shown. The probable date would be around the early to mid-1960s. The large cup being handed over would possibly have been displayed in the canteen, while the smaller cups would have been handed out to the individual members of the fire-fighting team to keep. (W. Rawstron Collection)

Below: Perhaps readers could also help with identifying the men in this photograph? The pit is Salterford No.2, and the occasion is of the men using the baths for the first time. A good deal of research went into trying to date this picture, but to no avail. It is known that Salterford No.2 closed in December 1959. (W. Rawstron Collection)

Hapton Valley Colliery was always known as a 'family pit', where father worked alongside son, and brother alongside brother. To some extent this tradition remained until the closure of the pit – as seen here, where brothers Harold and Dave Durant pose besides the upcast shaft at the colliery, c.1978. However, with the closure of other local collieries, the men from those pits were transferred to pits such as Hapton Valley and the family connections were to some degree diminished. Management were well aware of the long-standing family connections at the pit, and would try and place family members together where possible. At the same time, where possible those from other pits would also be placed to work alongside each other. This was not out of some kind commitment on the part of management – it was of course to get maximum production out of the men. (Harold Durant)

Over the many decades of coalmining in North East Lancashire there have been many hundreds of fatal accidents in the mines, yet there are few memorials to these sad events. There is, of course, the memorial to the Hapton Valley Colliery Disaster just inside the entrance to the Municipal Cemetery, and there is a memorial in the parish church of St Peter's at Burnley that records the death of a member of the Landless family at the Little Harwood Colliery in 1819. Inside St James's Church at Altham there is also a memorial to those who perished in the Moorfield Colliery Disaster in 1883. It was while researching this latter disaster that I came across a lone memorial at St James's. It reads: 'In loving memory of George Cronkshaw who was killed at the Habergham Colliery, 9th. December 1872 in his 48th. Year.' George perished at the Habergham Colliery, or Cheapside Pit as it was known locally, while the pit was still being sunk. The banksman at the pit, George had run a tub of bricks towards the shaft, thinking the gate was still down, but it wasn't, and the tub went over the edge of the 240yd-deep pit taking George with it. He was killed along with two of the shaft sinkers at the bottom of the pit, John Salkeld and Richard Starkie. Fifteen children were orphaned, and three widows were left to mourn their loss – just one instance that showed the dangers in mining coal.

Six
Sporting Days

Miners weren't just burly beer-drinking coal hewers, many were great sports players, and contrary to popular belief, it wasn't all greyhounds and racing pigeons. Each pit had its own football team, darts team, snooker team and cricket team, each as competitive as the next. During the war, one team captain commented that nobody could beat the pit lads when it came to football. He was soon shot down however, when someone stated, 'That's because every bugger else is in the army.' Miners, of course, were exempt from being called up for the war, as coal was needed to keep industry going. During the early days, the games were played on the local recreation grounds (recs) such as Whittlefield, Clifton and Rakehead – none of the glamour of grass, they played on cinders or ash-covered 'pitches'. Rough on the knees during tackles! Whittlefield Recreation Ground was ideal, the steep embankment running down to the 'pitch' provided an excellent viewing platform for the spectators. The author remembers as a child being raised on his granddad's shoulders at Whittlefield 'rec' to see his uncle, Terry Pickard, play. I was somewhat bewildered, as the fog that day was a real 'pea souper' and nobody could see more than a yard in front! Here, we see the Thorny Bank Colliery football team of 1957. Not all the men are known, but there is enough information to revive a few memories, I'll bet. From right: Noel Neely; -?-; Jimmy Gifford; -?-; Raymond Hope; -?-; Bill Abrams; Albert Lamamy, Bernard Morton; -?-;-?-;-?-.

Here we see one of the earliest pictures of a colliery football team, as shown on a 'Photocard' from Ardath Cigarettes, depicting the Hapton Valley Colliery Football Team, member of the Burnley and District Works Sports League. During the season 1935/36 they won the Victoria Hospital Cup, as well as on a previous occasion in 1925. They were also 'A' Division Champions in 1933/34. Back row, from left: J. Fletcher; J. McManus; J. Varley; W. Wren; R. Bullen; J. Taylor; T. Davenport. Front row, from left: J. Farrow; D. Warne; L. Davenport (captain); W. Cole; B. Whitaker; W. Bibby. (Archer Lee)

Opposite, above: Another 'Photocard' from Ardath Cigarettes featuring the Bank Hall Colliery Football Team, c.1935. The team were members of the Burnley and District Works Sports Welfare Association, and were champions in 1931/32, 1932/33 and 1935/36. They also won the Hospital Cup Final in 1935. For fifteen months they were undefeated in any competition, winning the Greenhalgh Cup and reaching the semi-final in the 1935/36 Season. Back row, from left: H. Smith; D. Forrest; J. Entwhistle; W. Walsh; J. Horner, T. Gill (secretary); D. Waddington. Front row, from left: P. Shorrocks, C. Burrows, J. Hawke, W. Wilkinson (captain); J. Pepper; W. Gilbert. Seated on the ground is the mascot. (Archer Lee)

Opposite, below: Pictured here at the last match for the Hospital Cup before its suspension is the Hapton Valley Colliery Football Team, who were beaten by Bank Hall Colliery 'A' Team 1-0. This was the last time any of the local pit teams ever played in this cup. The Hospital Cup, dating from 1884, is believed to be the oldest football competition cup in the world. It was restarted again in 1987 and is still being played for today. Back row, from left: A. Gregory (trainer); K. Tattersall; John Barker; B. O'Hara; B. Crossley; T. Kendall; W. Moore. Front row, from left: B. McDonald; Terry Pickard; L. McDevitt; Martin Feeney; P. Ramsbottom. Hapton Valley won the Hospital Cup in 1925 when they scored 2-0 against Haythornthwaites (Mill). They won the Cup again in 1936 when they drew 2-2 against Towneley Fireclay, beating them in the replay 2-1. (Archer Lee)

Hapton Valley Colliery, Works League Champions, 1952-53. Back row, from left: T. Greenwood; Frank Heywood; Alan Harvey; Dick Bentley. Middle row, from left: John Wilde; Brian Hoy; John Barker; Joe Walsh; Derek Redford; Henry Lee. Front row, from left: Harold Warne, manager; Albert Lee; Jim Hurley; Archer Lee; Lol McDevett; Neil Hurley; Norman Stowell. (Archer Lee)

Opposite, above: Clifton Colliery also had a decent football team – they were Burnley and District Works Sports Association League Champions, 1951-52, and finalists in the Greenhalgh Cup, 1952. The pit was closed down three years later. Back row, from left: H. Jenkins; R. Johnson; G. Chorlton; A. Barker; H. Veevers. Middle row, from left: S. Coulthurst; R. Whittaker; A. Simpson; J. Calvert; L. Smith; A. Lawson; W. Wiggin (trainer). Front row, from left: A. Tomlinson (president and manager); F. Wade; R. Farnworth; W. Robinson (captain); H. Allen; A. Whittaker; M. Thornber (secretary).

Opposite, below: An historic occasion: the Bank Hall 'A' team along with representatives of the Bank Hall 'B' team displaying the Hospital Cup and the No.4 Area Collieries Knockout Shield. The date is 1952, when Bank Hall 'A' team beat Bank Hall 'B' team in the Hospital Cup. Never before or since had both finalists playing for this coveted trophy been from the same works. The 'A' team was formed from the colliery workshops and the 'B' team from the pit bottom lads. Bank Hall Colliery held a proud record as far as the Hospital Cup went, winning it in 1904, 1935, 1952, 1953 and 1958. (Archer Lee)

When twenty-one-year-old cost clerk Derek Carter decided to form a Copy Colliery football team, he really took on a challenge. Although the pit employed around 170 men, only a handful of them actually played football. A team was formed, however, and in 1963, in only its second season, the team won the Colonel Bolton Trophy by defeating Bank Hall's 'A' team 2-5 on 31 March. This brought them into the Burnley Area Championship, which took them to the Divisional semi-final against Chanters Colliery at Atherton. The game was played at the Bank Hall Miners' Welfare Ground on 21 April 1963. The final score was Chanters 4, Copy 2, a credible score for Copy, the team from the little pit who had taken on the giant killers. The team are, standing, from left: Philip Wilding; Gerald Locke; George Rawstron; Brian Wiggin. Front row, from left: Ronnie Bannister; Derek Carter (captain); Alan Dickinson; Jimmy Pollard; Brian Gildert; Tommy Gildert; Barry Fothersgill. (Derek Carter)

Moving on to other sports, we see here the Bank Hall Colliery Sports and Social Club Champions, 'A' Division, 1947-48, who won the Burnley Works Association Snooker League that year. Back row, from left: G. Waddington; R. Duckworth; S. Reynolds. Middle row, from left: A. Finney; W. Walsh (secretary). Front row, from left: S. Hind; W.E. Rawstron (vice president); F. Pepper (captain). (Ian Rawstron)

Having mentioned Bank Hall's snooker team, I'd better also include one of Hapton Valley's snooker teams – to avoid screams of prejudice! Here is the 1953 winning team. Standing: Thomas Greenwood. Front row, from left: Dick Bernard; Frank Haywood; William Herbert Large; Ben Rushton; Herbert Large.

Besides having snooker, football and cricket teams, Bank Hall Colliery also had a decent swimming team, as seen here c.1941. Back row, from left: Albert Dugan; Albert Whittaker; Frank Hargreaves (swimmer); Fred Large; Johnny Walker; David Forrest. Middle row, from left: Edward Parkinson; Arthur Dobson. Front row, from left: Harry Broxup; Harry Adcroft; S. Marsden; Martin Ashworth.

Seven
The Private Coal Mines

When Britain's coal mines were Nationalised in January 1947, the new NCB had the option of granting licenses to persons who wished to operate their own private mines. A year later, in 1948, there was just one private coal mine in the area covered by this book – the Blackclough Lower Mountain Mine. This colliery, worked by the Deerplay Colliery Co., employed thirty-three men underground and eight surface workers. The pit was worked out by 1951. There were, between this date and 1960, a number of small private pits such as the Inchfield Moor Colliery on the Bacup Road at Todmorden. Although this pit did mine coal, it was mainly a fireclay mine worked by Temperley's for their fireclay products. The Merrills Head Colliery was begun around 1960 by John Simpson Little of Merrills Head Farm on the Long Causeway. Mr Little began his career in coalmining around 1950 at the Copy Colliery, working at the pit for around ten years. When he first began mining in his own right, taking the coal near its outcrop on his own land, the manager of the Copy Pit, one of the Jobling's, blew up Mr Little's mine with dynamite. All the mineral rights in Britain belonged to the NCB following Nationalisation, and Mr Little was actually taken to court for the offence of 'Maliciously removing minerals and coal belonging to the NCB'. The magistrates saw the triviality in a person digging on his own land, and dismissed the case. Having got his deputy's certificate, Mr Little was able to apply for a licence for a private mine, but had to wait six years before it was eventually granted. 'The Coal Board often saw us, the private mine owner, as a fly in the ointment,' he said. The first of the five Merrills Head Colliery drift pits was begun around 1960. Here we see a miner emerging from the drift entrance to the Merrills Head Colliery. (John Simpson Little)

The Merrills Head Collieries mined the coal from isolated beds of the Arley Seam, the best seam in the Burnley Coalfield. With good detective work, Mr Little was able to work out where these beds of coal had been left, following extensive opencast operations in the area and from earlier mining from places such as Railway Pit in the valley below and the numerous small pits that worked along the lines of outcrop. Here a miner named Andrew (last name unknown) brings out a tub from the Merrills Head Colliery around 1985. The tubs ran on L section rails, cheaper and easier to come by than the traditional 'railway' type rails. Sturdy timbers needed for the gantry above the hopper near the drift mouthing were old flooring joists from places such as the derelict Burnley Mills, some as thick as 18in by 12in. For the wheels of the tubs, the brake disks from old Ford cars were used – it had to be a particular make of Ford however, one which incorporated ball bearings. (John Simpson Little)

Output at the Merrill Head Mines averaged around 100 to 150 tons per week, going mainly for household use. Those customers who wanted a bag or two of household coal simply drove up to the Merrills Head Farm and asked for the quantity they wanted. Like all youths, the lads at Merrills Head often found time for a bit of fun; here one of the lads receives his 'comeuppances' from the other lads at the colliery in the pit top cabin. Left to right are: Rodney Mitchell; Graham Topping; Bernard Rathall; Michael Salmon and Billy Clayton. (John Simpson Little)

The Merrills Head Collieries were somewhat scattered about; one was near the entrance gates to the farm itself, one in a field on the Burnley side of the farm, and one near local historian Titus Thornber's Middle Pasture Farm. Near the Merrills Head Farm itself, the pit was well equipped as far as the surface arrangements were concerned for the distribution of output, with hoppers, weigh scales and bagging facilities. Here, Graham Topping can be seen bagging coal at the Merrills Head Colliery, 6 April 1976. (John Simpson Little)

In 1980, the Merrills Head Colliery employed three men underground and one man on the surface. The pits at this time were numbered one and five. Here, Bernard Rathall is seen in charge of the surface winch at the colliery. Haulage engines were adapted from any old diesel engine which happened to be around at the time, and were modified for use at the pit – add a piece here, take a piece away from there, wrap some wire rope around a drum, and there you are – a colliery haulage engine. (John Simpson Little)

Although the day-to-day running of any private mine might seem rudimentary by modern pit standards, the pits were nevertheless inspected regularly by the Mines Inspector for the area. Any defects or unsafe conditions had to be rectified immediately or the pit could be shut down at a moments notice. The Arley Seam at Merrills Head was by all accounts a good seam to work in, being about 3ft high. Rodney Mitchell is seen here at the coalface at Merrills Head Colliery around the mid-1970s. (John Simpson Little)

Eight
The End of Coalmining in North East Lancashire

Closures came with alarming quickness in the Burnley area following Nationalisation, and the first pits to go were the Towneley and Dyneley Collieries, followed by the Fence Drift Mines, which in all fairness were only short-term workings. The last tub of coal at the Clifton Colliery was raised on 30 December 1955, thus ending almost eighty years of coal production. In the last full year of production, the Clifton Colliery turned out 39,483 tons of coal and employed 219 men, many of whom were found employment at other local pits when Clifton closed. Clifton, like many other local pits, was very much a 'family pit' and there must have been at least some twinge of sadness when the pit finally closed. Here, F.C. Bowers receives his last wage from wage clerk, S. Coulthurst, after working over thirty years in the coalmining industry. The latter, it will be remembered, was an active member of the colliery football team. I'm afraid I don't know the name of the gentleman in the background of the office having a fag! Although Clifton had been shut down, the surface buildings remained for many years afterwards, in fact a family and two young daughters lived in the manager's house until the mid to late 1970s. With the coming of the M65 motorway the whole area of the former colliery was covered over with spoil from the Whittlefield cutting of the motorway. There are now no remains of Clifton Colliery, the site of which has now been 'beautified' by being renamed Clifton Heights. (W.E. Rawstron Collection)

The end finally came for the Reedley Colliery, Burnley, on 20 May 1960, a closure caused through uneconomic workings. Since the sinking of the colliery in 1879, the pit worked the Arley Mine of excellent household coal. This seam had been worked out from around 1935, after which the Dandy Mine was worked, until this too was worked out. Here the men on the last shift at Reedley Colliery, always known locally as Barden Pit, walk away from the pit shaft for the last time – many were found employment at other local pits, notably at Hapton Valley Colliery. The men are, from left: Jimmy Ormerod; Fred Harrison; Arthur Chadwick; Joe Brown; Wilkinson Spensor; Clifford Torkington; John Pollard. Another employee, Fred Harrison, had been employed at the coalface at Reedley since 1919, and before that he had worked at both the Towneley and Bank Hall Pits. Arnold Mort, sixty-four, the lamp-man at the pit, decided to call it a day. During his fifty-one years service in the pits he had seen three other collieries close and had been on the staff at Clifton, Habergham and Higham Pits, all now gone. The overall output at the colliery was well over 40cwt per manshift, as good as anything in the Lancashire area, and it produced 3,000 tons per week when working at its best. The colliery also had a proud record in the field of sport. Its football team won the Hospital Cup three times in succession in 1920, 1921 and 1922. They were also the runners-up in the final of the Hospital Cup in 1907 against Duckett's Works. In 1960 they won the area football knockout shield, the area snooker competition, and the Works Sports Cricket League. In 1959, the darts team won the local works league competition and the NCB knockout competition. The site of Reedley Colliery today has been landscaped, although one part is now used as a Quick Mix concrete works. (Burnley Borough Council, Towneley Hall Art Gallery and Museum)

Time was running out for the Fir Trees Colliery backshift team when this photograph was taken in 1965 – the colliery was closed down on 11 March 1966. Most of the sixty-eight miners employed at the pit were found alternative work at Hapton Valley Colliery. Back row, from left: Fred Atkinson; Wilf Derbyshire; Jimmy Bennett; Peter Lofthouse; Joe Bennett. Front row, from left: Joe Brown; Ned Kenyon; Fred Hardacre. There are no remains of the old Fir Trees Colliery; the only reminder is a solitary electricity pole, which still bears a metal plate stating 'Fir Trees Colliery'. (Joe Bennett)

How the mighty fall! The tumbling headgear, the once proud symbol of a great industry, crumbles to the ground at Bank Hall Colliery in 1971, marking the end of over one hundred years of coalmining history. Thirty years on, the area that used to accommodate Burnley's largest colliery is now a large and pleasant recreation space. Near the Eastern Avenue entrance to the former colliery there now stands a memorial to coalmining in Burnley. The Bank Hall Colliery was finally closed down because of frequent ignitions on the coalface and the fear of an explosion – others might say that the fact that the colliery lost almost £1 million in its last year was the deciding factor, along with the political and economics changes at that time. During the pit's last full year of production, the 722 men employed at the pit only managed to raise 170,456 tons of coal, equal to just over one ton per man per shift. Compare that with Hapton Valley Colliery, which on closure was producing over three and a half tons per man per shift. Many of the men at Bank Hall decided to call it a day and took the redundancy money, others were found employment in other parts of Lancashire and moved on like industrial nomads to the Agecroft, Point of Ayr and Parkside Collieries, until they too were closed down.

This picture of the lads on the M6 coalface is supposedly of the very last shift worked at Hapton Valley Colliery. Back row, from left: Jimmy ?; P. Pounder; Frank ?; Dave Durant; David Pickup; Alan Etherington; Peter Bradley; Ken Holland; Len Macintosh; Len Fallows. Front row, from left: Peter Monroe; Tony Pickup; David Lord; Frank Greenhalgh; Joe Cheetham; his brother Tony Cheetham; Norman Holden. Some of the men were retained to salvage the pit but, soon after, the colliery headgear was demolished. Most of the surface buildings are still in place at the former colliery, which is now used by a skip hire place. Soon afterwards the miles of underground tunnels flooded to the roof and the water burst out near the old surface drift, spewing pollution into Green Brook below. When I enquired about this pollution to the Water and Environment Authorities, they both said they had no control over mine-water pollution and that it was down to the Coal Authority. Private mine owners are obliged to restore any land where mining has taken place to its former condition – it's a pity the same rule doesn't apply to the Coal Authority.

Footnote

From the date the coal mines of Great Britain were Nationalised, 1 January 1947, to March 1984, a period of less than forty years, sixty-four Lancashire coal mines were closed down with a loss of 32,632 jobs. Hapton Valley was the last deep coal mine in North East Lancashire to close down in 1982, Parkside Colliery at Newton-le-Willows was the last deep coal mine in Lancashire when it closed down in 1993.